AF388952

Study of the Soil Water Movement in Irrigated Agriculture

Study of the Soil Water Movement in Irrigated Agriculture

Editors

George Kargas
Petros Kerkides
Paraskevi Londra

MDPI • Basel • Beijing • Wuhan • Barcelona • Belgrade • Manchester • Tokyo • Cluj • Tianjin

Editors

George Kargas
Natural Resources and
Agricultural Engineering
Agricultural University
of Athens
Athens
Greece

Petros Kerkides
Natural Resources and
Agricultural Engineering
Agricultural University
of Athens
Athens
Greece

Paraskevi Londra
Natural Resources and
Agricultural Engineering
Agricultural University
of Athens
Athens
Greece

Editorial Office
MDPI
St. Alban-Anlage 66
4052 Basel, Switzerland

This is a reprint of articles from the Special Issue published online in the open access journal *Water* (ISSN 2073-4441) (available at: www.mdpi.com/journal/water/special_issues/Soil_Water_Movement_Agriculture).

For citation purposes, cite each article independently as indicated on the article page online and as indicated below:

LastName, A.A.; LastName, B.B.; LastName, C.C. Article Title. *Journal Name* **Year**, *Volume Number*, Page Range.

ISBN 978-3-0365-6196-7 (Hbk)
ISBN 978-3-0365-6195-0 (PDF)

Cover image courtesy of Paraskevi Londra

© 2023 by the authors. Articles in this book are Open Access and distributed under the Creative Commons Attribution (CC BY) license, which allows users to download, copy and build upon published articles, as long as the author and publisher are properly credited, which ensures maximum dissemination and a wider impact of our publications.

The book as a whole is distributed by MDPI under the terms and conditions of the Creative Commons license CC BY-NC-ND.

Contents

About the Editors

George Kargas

Dr. George Kargas is Professor at the Agricultural Hydraulics Laboratory of the Department of Natural Resources and Agricultural Engineering, of the Agricultural University of Athens (Greece). His first degree (equivalent to MSc) was in Agricultural Engineering and his PhD in Soil Physics. His research interests include: Soil Water Movement in the Unsaturated and Saturated Zone; Hysteresis in Pore Water; Determination and Evaluation of Soil and Substrates Hydrological and Physical Properties; Irrigation and Drainage, Investigation of Soil Water and Solute Regime in Irrigated Agriculture by the Use of Dielectric Devices. He has published more than 150 peer-reviewed articles in international scientific journals, in conference proceedings, specific chapters in relevant books and other articles. He has also served as a reviewer for many peer-reviewed international journals. He is Guest Editor of the journal *Water*.

Petros Kerkides

Dr. Petros Kerkides is Professor Emeritus of The Agricultural University Athens (Greece). He has served as a member of the Academic Governing Committee of The Technological University of Cyprus in Limassol, at its first period of establishment, and also as a member of the Academic Governing Committee, for two successive 4-year sessions, of the Open University, Nicosia Cyprus. His first degree is in Physics from the National Kapodistrian University of Athens, Greece and his PhD in Solid State Physics from Salford University, U.K. Both degrees were obtained under national grants, after successful exams. His research interests include: Solid state physics, soil water movement, irrigation and drainage in agricultural fields, evapotranspiration, solute transport, etc. He has published a large number of peer-reviewed articles in international scientific journals, in conference proceedings, specific chapters in relevant books and other articles. He has served as a reviewer for a number of international scientific journals, and as scientific supervisor for a number of PhD theses. He is a guest Editor of the MDPI journal *Water*.

Paraskevi Londra

Dr. Paraskevi A. Londra is Assistant Professor at the Agricultural Hydraulics Laboratory of the Department of Natural Resources and Agricultural Engineering, of the Agricultural University of Athens (Greece). Her first degree (equivalent to MSc) was in Agricultural Engineering, and her PhD was in Soil Physics and Irrigation. She has also an MSc degree in Hydrology and Water Resources Environmental Management. Her research interests include: Physical-Hydraulic Properties of Soil and Substrates, Vadose Zone Hydrology, Irrigation Management, Plant Soil-Water Interaction, Water Resources Management. She has published more than 120 peer-reviewed articles in international scientific journals, in conference proceedings, specific chapters in relevant books and other articles. She serves as a reviewer in many peer-reviewed international journals. She is a member of the Editorial Board of the journal *Agriculture and Forestry* and Guest Editor of the journal *Water*.

Article

Investigation of the Flux–Concentration Relation for Horizontal Flow in Soils

George Kargas, Paraskevi Londra * and Petros Kerkides

Department of Natural Resources Management and Agricultural Engineering, Agricultural University of Athens, 75 Iera Odos, 11855 Athens, Greece; kargas@aua.gr (G.K.); kerkides@aua.gr (P.K.)
* Correspondence: v.londra@aua.gr; Tel.: +30-210-529-4069

Received: 6 October 2019; Accepted: 19 November 2019; Published: 21 November 2019

Abstract: The objective of the present work is to investigate the flux–concentration ($F(\Theta)$) relation, where Θ is the normalized soil volumetric water content for the case of one-dimensional horizontal flow, subject to constant concentration conditions. More specifically, the possibility of describing $F(\Theta)$ by an equation of the form $F(\Theta) = 1 - (1 - \Theta)^{p+1}$ is examined. Parameter p is estimated from curve-fitting of the experimentally obtained $\lambda(\Theta)$ data to an analytic expression of the form $(1 - \Theta)^p$ where λ is the well-known Boltzmann transformation $\lambda = xt^{-0.5}$ (x = distance, t = time). The results show that the equation of $(1 - \Theta)^p$ form can satisfactorily describe the $\lambda(\Theta)$ relation for the four porous media tested. The proposed $F(\Theta)$ function was compared with the limiting $F(\Theta)$ function for linear and Green–Ampt soils and to the actual $F(\Theta)$ function. From the results, it was shown that the proposed $F(\Theta)$ function gave reasonably accurate results in all cases. Moreover, the analytical expression of the soil water diffusivity ($D(\Theta)$) function, as it was obtained by using the equation for $\lambda(\Theta)$ of the form $(1 - \Theta)^p$, appears to be very close to the experimental $D(\Theta)$ data (root mean square error (RMSE) = 0.593 m²min⁻¹).

Keywords: horizontal infiltration; sorptivity; soil moisture profile

1. Introduction

For a rational and productive irrigation water application, one needs to know well how water infiltrates into the soil, and how the soil moisture profiles evolve. This can be achieved by solving the appropriate Richards' equation, which is feasible only when major soil hydraulic properties are known, together with the initial and boundary conditions imposed in each case.

To solve the unsaturated flow equation, many procedures have been proposed, e.g., analytical, semi analytical, finite differences, and finite elements. Philip [1] introduced the flux–concentration ($F(\Theta)$) relation in order to develop a quasi-analytical technique for solving the unsaturated flow equation. The technique is rather simple and gives a holistic view of the phenomenon [1,2].

$F(\Theta)$ expresses the relation of soil water flux–density with the normalized volumetric soil water content, and can be used in an attempt to elucidate further the flow phenomena in unsaturated soils, both in the horizontal (absorption, as well as desorption) and in the vertical cases. The form of the F for the case of horizontal absorption and the normalized volumetric soil water content is given by the following equations:

$$F = \frac{q}{q_0}, \quad \Theta = \frac{\theta - \theta_{in}}{\theta_0 - \theta_{in}} \tag{1}$$

where θ_{in} is the initial volumetric soil water content of the soil column, considered uniform throughout the column; θ_0 is the volumetric soil water content at the soil surface ($x = 0$); q is the soil water flux density at a position x where soil water content is θ; and q_0 is the soil water flux density at $x = 0$, the soil surface, where water is absorbed under constant concentration conditions with $\theta = \theta_0 = \theta_s$

(θ_s = volumetric soil moisture at saturation) when the soil water pressure head is $H = H_0 = 0$ cm column of water.

The values of F and Θ lie in the range of 0 to 1. A zero value for F and Θ when q = 0 is far enough from the soil infiltration surface, where $\theta = \theta_{in}$; the value is 1 at the soil surface, where $q = q_0$ and $\theta = \theta_0$ [1]. In general, the F is a function of θ, θ_{in}, θ_0, and time (t) and it is, in most cases, unknown a priori [3].

For the prediction of the flow regime in porous media, using the flux–concentration relationship, either an iterative procedure should be applied, or accurate estimates of $F(\Theta)$ where feasible are used [2].

A number of researchers have proposed analytical expressions for the $F(\Theta)$ relationship to solve the flow equation for the case of the horizontal absorption under constant pressure head conditions, applied at the soil surface ($x = 0$).

Smiles et al. [4] proposed an empirical expression to describe the measured $F(\Theta)$ of a fine sand soil:

$$F(\Theta) = 1 - (1 - \Theta)^{1.19}. \tag{2}$$

Vauclin and Haverkampt [5] proposed the following simple expression

$$F(\Theta) = \frac{2\Theta}{\Theta + 1}, \tag{3}$$

while Evangelides et al. [6] proposed the empirical expression

$$F(\Theta) = 2\Theta - \left[1 - (1 - \Theta)^m\right] \tag{4}$$

with the value of parameter m lying in the range 0.718–0.867, and the value $m = 0.8$ considered as the most appropriate, which is very close to the mean value.

Recently, Ma et al. [7] proposed the expression

$$F(\Theta) = \Theta\left[1 + a_2\left(1 - \Theta^{\frac{1}{a_2}}\right)\right] \tag{5}$$

with the parameter a_2 being a function of the initial soil water saturation and the soil pore structure index that reflects the shape of the soil water retention curve.

From the abovementioned expressions, those where F is a function of Θ with a parameter for soil properties (Equations (4) and (5)), compared to those where F is a function of Θ without any parameters (Equations (2) and (3)), are more flexible and can simulate $F(\Theta)$ relationships with higher accuracy for a wide range of texture soils [7].

Philip [1] investigated the behavior of F for two distinct cases of soils, characterized by their $D(\theta)$ functions—i.e., soils, where their $D(\theta)$ function is a delta function (the so called Green–Ampt soils, where the moisture profiles advance step-wise), as well as soils whose $D(\theta)$ function is constant, independent of moisture content θ (the so called linear soils). In real soils, the $F(\Theta)$ relationship will lie in between the Green–Ampt and linear soils [1].

Clothier et al. [8] investigated the possibility of estimating the expression $\lambda(\Theta)$, where λ is the Boltzmann transformation $\lambda(\theta) = xt^{-0.5}$ for each soil from the experimental data of a horizontal absorption experiment, according to the equation below (Philip [9], Table 1, no. 2):

$$\lambda(\Theta) = \lambda_i(1 - \Theta)^p, \tag{6}$$

where λ_i is the maximum value of $\lambda(\Theta)$ (i.e., when $\Theta = 0$). Parameter λ_i considered to be a characteristic measure of the wetted region in a horizontal absorption experiment, and is influenced only by θ_{in} for every soil [10–12]; p is a fitting parameter.

Equation (6) has the advantage of an easy estimation of parameters λ_i and p; by a log-transformation, Equation (6) becomes linear, with the slope equal to p and transect equal to $\log\lambda i$ when the $\lambda(\Theta)$ is described by Equation (6). This procedure appears preferable to least squares fitting.

It is to be mentioned here that previous similar studies [6] could obtain the $F(\Theta)$ relationship using more complex expressions for it. In the present study, we use a simple, mono-parametric $F(\Theta)$ expression using a mono-parametric $\lambda(\Theta)$ expression, which adequately describes the experimental data [8].

In this context, the main objectives of this study are (1) to test the possibility of the mono-parametric $\lambda(\Theta)$ expression to describe the experimental data, as the equation $\lambda(\Theta) = \lambda_i(1 - \Theta)^p$ through the proper selection of a fitting parameter p; and (2) to investigate whether the experimental $F(\Theta)$ relations are described accurately enough from the analytical, mono-parametric function $F(\Theta) = 1 - (1 - \Theta)^{p+1}$ for a wide range of soil types, as well as finding out the physical meaning of the parameter p and the range of its values.

2. Theory

Applying Darcy's law and the mass conservation principle for the one-dimensional horizontal infiltration case, one easily gets the partial differential equation

$$\frac{\partial \theta}{\partial t} = \frac{\partial}{\partial x}\left[D(\theta)\frac{\partial \theta}{\partial x}\right], \tag{7}$$

where x is the horizontal axis, t the time, and $D(\theta)$ the soil water diffusivity. The solution of Equation (7) under the following initial and boundary conditions:

(a) $t = 0, x > 0, \theta = \theta_{in}$
(b) $t > 0, x = 0, \theta = \theta_0$
(c) $t > 0, x \to \infty, \theta = \theta_{in}$

could be obtained in terms of the Boltzmann transformation $\lambda(\theta) = xt^{-0.5}$, considered a function of θ only as

$$\int_{\theta_{in}}^{\theta} \lambda(\theta)d\theta = -2D(\theta)\frac{d\theta}{d\lambda}, \tag{8}$$

following the initial and boundary conditions $\theta = \theta_{in}, \lambda \to \infty$, and $\theta = \theta_0, \lambda = 0$.

Introducing the Boltzmann transformation function $\lambda(\theta) = xt^{-0.5}$ in Darcy's law, one gets

$$q = -K(\theta)\frac{dH}{dx} = -D(\theta)\frac{d\theta}{d\lambda}\frac{\partial \lambda}{\partial x} = -t^{-0.5}D(\theta)\frac{d\theta}{d\lambda}, \tag{9}$$

and the combination of Equations (1), (8), and (9) gives the following expression for F

$$F = \frac{\int_{\theta_{in}}^{\theta} \lambda(\theta)d\theta}{\int_{\theta_{in}}^{\theta_0} \lambda(\theta)d\theta}, \tag{10}$$

From the abovementioned equation, it is obvious that function F, for the case of the horizontal absorption, is independent of time and depends on the values of the volumetric soil water, θ, θ_{in} and θ_0.

When the $D(\theta)$ function is a delta function, the flux concentration relationship F for the horizontal absorption is given by

$$F = \Theta, \tag{11}$$

while for the case where the $D(\theta)$ function is constant, the flux concentration relationship F is given by

$$F = \exp\left[-(\text{inverfc}(\Theta))^2\right], \tag{12}$$

White et al. [13] have shown that Equation (12) could be reasonably approximated by the expression

$$F = \sin\left[\left(\frac{\pi}{2}\right)\Theta^{\frac{\pi}{4}}\right].$$

(13)

It is obvious from Equations (8) and (10) that the $F(\Theta)$ relationship depends on the form of $D(\theta)$.

From Equation (6), the soil sorptivity S_c function could be estimated by analytical expression [8], since

$$S_c = (\theta_0 - \theta_{in})\int_0^1 \lambda_i(1 - \Theta)^p d\Theta = \frac{(\theta_0 - \theta_{in})\lambda_i}{p + 1},$$

(14)

From the combination of Equations (6), (9), and (10), it could be shown that the $F(\Theta)$ relationship could be estimated analytically, according to the mono-parametrical analytical expression

$$F(\Theta) = 1 - (1 - \Theta)^{p+1}.$$

(15)

A similar expression for $F(\Theta)$ was presented by Smiles et al. [4] in a horizontal absorption experiment in sand, where the parameter p had a constant value $p = 0.19$.

Applying the Bruce and Klute [14] method, the soil water diffusivity function $D(\Theta)$ can be estimated according to the expression

$$D(\Theta) = -\frac{\int_0^\Theta \lambda(\Theta)d\Theta}{2\frac{d\Theta}{d\lambda}}.$$

(16)

In this respect, it is rather easy, by introducing the $\lambda(\Theta)$ expression given in Equation (6), to get an analytical expression for $D(\Theta)$ ([8]), as

$$D(\Theta) = \frac{p(p + 1)S^2\left[(1 - \Theta)^{p-1} - (1 - \Theta)^{2p}\right]}{2(\theta_0 - \theta_{in})^2}.$$

(17)

In practice, this means that for soils with a $\lambda(\Theta)$ expression, such as the one given by Equation (6), their diffusivity function $D(\Theta)$ should be described analytically by Equation (17), and the flux concentration relationship $F(\Theta)$ by Equation (15).

Table 1. The values of the soil characteristics θ_0, θ_{in}, ρ_φ, and S, together with the values of the parameters λ_i, p, and S_c for each soil examined. The value of the parameter p was estimated from Equation (6). The value of sorptivity S was obtained directly from the experimental $I(t)$ data (Equation (18)), while S_c comes from Equation (14). RE denotes the absolute relative error between actual S and estimated S_c values for each soil examined.

Porous Media	θ_0 (m³ m⁻³)	θ_{in} (m³ m⁻³)	ρ_φ (g cm⁻³) [1]	λ_i (cm min⁻⁰·⁵)	p	S (cm min⁻⁰·⁵)	S_c (cm min⁻⁰·⁵)	RE (%)
Sand (S)	0.284	0.06	-	12.210	0.35	1.930	1.980	2.59
Sandy loam (SL)	0.418	0.015	1.41	3.510	0.12	1.300	1.26	3.08
Loam (L)	0.465	0.022	1.12	1.545	0.1	0.644	0.634	1.55
Silty clay loam (SiCL)	0.511	0.028	1.22	0.656	0.18	0.309	0.268	13.27

[1] Dry soil bulk density value for sand soil is not referred to by Poulovassilis [15].

3. Materials and Methods

Horizontal infiltration experiments were conducted (Scheme 1) in three porous media with different soil textures: a sandy loam (SL) (13.2% clay, 8% silt, 78.8% sand), a loam (L) (20% clay, 38% silt, 42% sand), and a silty clay loam (SiCL) (36.5% clay, 52% silt, 11.5% sand). The respective soils were air-dried, ground, and passed through a 2 mm sieve. The soil samples were uniformly packed in a rectangular column 5 cm in width, 2 cm in height, and 50 cm long. The upper soil surface was open to the air, thus securing that the soil air pressure at the soil pores was at atmospheric pressure

(Scheme 1). The experimental data from horizontal absorption experiments conducted in sand (S) by Poulovassilis [15] were also used.

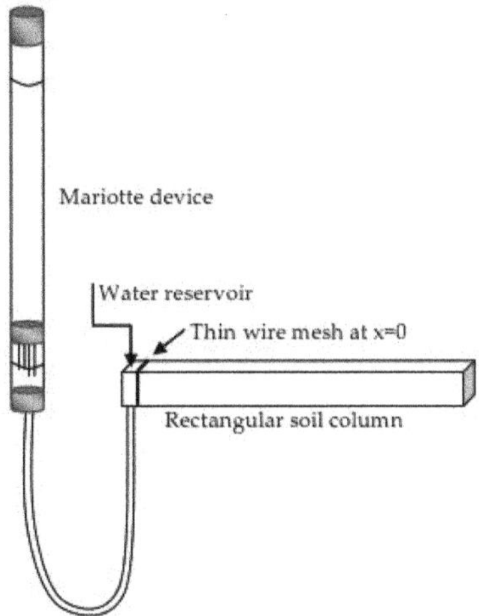

Scheme 1. Experimental device.

The volumetric soil water $\theta_j(x_j, t^*)$ profile, after a time period t^* was elapsed from the beginning of the experiment, was determined by cutting, as quickly as possible, the soil column into small rectangular pieces. Specifically, the soil column was sectioned in 0.01 m increments, and the volumetric water content of each rectangular piece θ_j was determined by using the gravimetric water content and the dry soil bulk density (ρ_φ). The θ_j value corresponds to the center of the soil samples, at distance x_j from the soil infiltration surface.

The initial values of θ and θ_{in} were those associated to the air-dry soil, and the condition of the soil infiltration surface was that of a constant pressure head H value ($H = 0$ at $x = 0$, with $x = 0$ denoting the soil infiltration surface). The zero value of H was maintained by a Mariotte device (Scheme 1). A thin wire mesh at $x = 0$ was installed to keep the soil at rest, and also provide the least possible resistance to the soil water entry.

The time duration of these experiments varied according to the soil type—i.e., smaller values for coarse-textured soils and larger values for fine-textured soils. For the sand (S), it was 10.4 min [15], for the SL it was 194 min, for the L it was 251 min, and for the SiCL it was 1630 min. During the experimental process, a continuous monitoring of the water volume entering the column was obtained, thus allowing the cumulative infiltration $I(t)$ experimental curve to be determined. The soil sorptivity S is immediately available from the well-known relationship [16]

$$S = \frac{I}{t^{0.5}},\tag{18}$$

4. Results and Discussion

In Table 1, some soil characteristics, such as θ_0, θ_{in}, ρ_φ, λ_i and S (Equation (18)), together with fitting parameter p and the value of S_c (Equation (14)), are shown for each porous media used in this study. It is easily noticeable that the values of S and λ_i depend strongly on the soil type, and they tend to decrease as soils become finer in texture. The same trend for the values S and λ_i for six different soils were presented from McBride and Horton [10].

In Figure 1, the $\lambda(\theta)$ profiles are presented, as these were obtained from the experimental $\theta(x,t)$ data and the well-known Boltzmann transformation ($\lambda(\theta) = xt^{-0.5}$), together with the fitted $\lambda(\theta)$ relationship (Equation (6)), after the fitting parameter p was properly selected. From the comparison, it is shown that Equation (6) is suitable for the description of the $\lambda(\theta)$ relationships for the horizontal absorption experiments. In each soil, the p-value was determined as the one where the root mean square error (RMSE) from a series of neighboring p-values was the least.

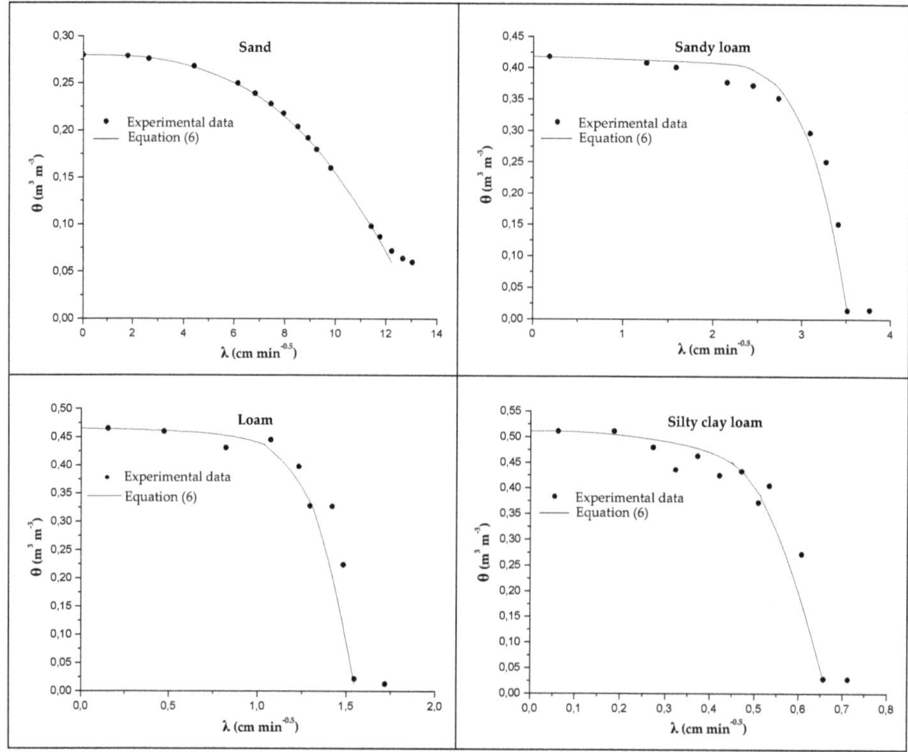

Figure 1. The $\lambda(\theta)$ relationships, as obtained directly from experimental $\theta(x,t)$ data for each soil, and the estimated values according to Equation (6).

One could argue, considering that the $\lambda(\Theta)$ relationship is unique for each soil, independent of the duration of the experiment, that the two parameters of Equation (6) (λ_i and p) do not represent simple fitting parameters, but are related to the soil's hydraulic properties. Some researchers insist that parameter λ_i may be considered as an index of the soil's hydraulic properties [10,11]. Moreover, Shao and Horton [11] correlated parameter p with the parameter n ($p = 1/n$) of the equation of van Genuchten [17], which describes the soil moisture retention curve.

In what follows, an investigation of the characteristics of parameter p will be carried out. According to Equation (14), the ratio $S/\lambda_i(\theta_0 - \theta_{in})$ is equal to $(p + 1)^{-1}$. For the Green–Ampt soils (D is a Dirac delta-function of θ), the ratio $S/\lambda_i(\theta_0 - \theta_{in})$ should be unit, thus p would be zero. Similarly, for the linear soils, the expression $(p + 1)^{-1}$ would be 0.31, and the value of p will be 2.23 [8]. Consequently, the values of p are related to the form of the diffusivity $D(\theta)$ function. In order to examine this relationship, Equation (6) is rewritten as in Equation (19):

$$\frac{\lambda}{\lambda_i} = (1 - \Theta)^p \rightarrow \Theta = 1 - \left(\frac{\lambda}{\lambda_i}\right)^{\frac{1}{p}}. \tag{19}$$

In Figure 2, the $\Theta(\lambda/\lambda_i)$ relationship is shown for various values of the parameter p ($0 < p < 2.23$). From Figure 2, it is shown that all the $\Theta(\lambda/\lambda_i)$ relations lie in the area with its borders defined by the values of the parameter p (i.e., $p \rightarrow 0$; $D(\theta)$ Dirac delta function) and $p = 2.23$ ($D(\theta)$ constant). From the investigation of the $\lambda(\theta)$ relations in this study, it is found that p-values fall in the range $0.1 < p < 0.4$. Also, Evangelides et al. [6] showed that p-values obtained using data from horizontal infiltration experiments in seven soils fall in the range $0.149 < p < 0.389$. In other words, the range of p is narrower than the range $0 \leq p \leq 2.23$.

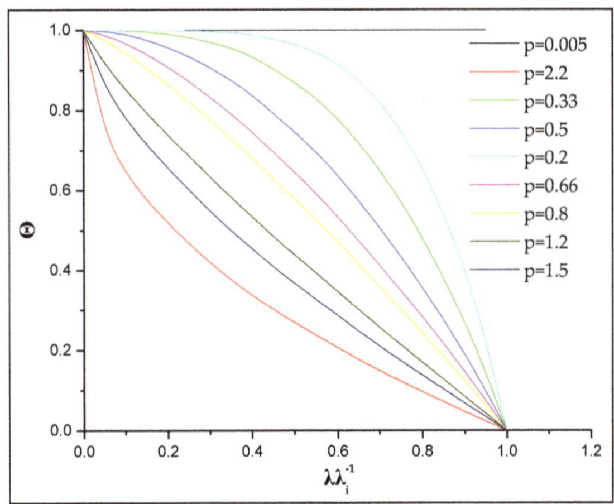

Figure 2. The relationship $\Theta(\lambda/\lambda i)$ for different values of the parameter p.

After the parameter p selection, using also the λ_i, θ_0, and θ_{in} values for each soil, a re-estimation of sorptivity S_c was performed using Equation (14). From the values of S and S_c presented in Table 1, S_c values are reasonably close to the experimental values of S ($S = I/t^{0.5}$) for three out of the four soils examined (absolute values of RE: 1.55% < RE < 3.08%). The relative error value for the SiCL soil appears to be rather high (13.27%). This could be attributed to the long time duration of the experiment (1630 min) and unavoidable soil water evaporation from the upper soil surface. In any case, the overall differences are small, and therefore one may consider that Equation (14) can lead to a quick and reliable way of estimating S from a set of horizontal absorption experimental data.

In Figure 3, the $D(\theta)$ data for the loamy soil (L) are shown. Closed circles denote data obtained by the Bruce and Klute [14] method (Equation (16)), while open circles are data obtained from Equation (17). The comparison of the above indicates that Equation (17) could reasonably describe the experimental $D(\theta)$ data. One should note that by using the analytical expression (Equation (17)), the problem of differentiating the experimental data $\frac{d\Theta}{d\lambda}$, in which there is scatter, is overcome. Moreover, the problem of estimating the slope $\frac{d\Theta}{d\lambda}$ near saturation, where the $\lambda(\Theta)$ relationship is almost parallel to the λ-axis, is overcome by the application of the analytical expression (Equation (17)). Similar results were also obtained for the other soils under present investigation.

In Figure 4, the $F(\Theta)$ relationship for the linear Equation (13) and the Green–Ampt soils (Equation (11)) is shown, together with that obtained according to Equation (15) for all soils examined. The experimental $F(\Theta)$ points for all soils were obtained by the application of Equation (10) and the measured value of the sorptivity S (Equation (18)). From the results, it is shown that the one-parameter Equation (15) gave practically the same values for the $F(\Theta)$ relationships as the ones obtained experimentally, and lies between the two limits (linear and Green–Ampt soils) in all cases that were examined.

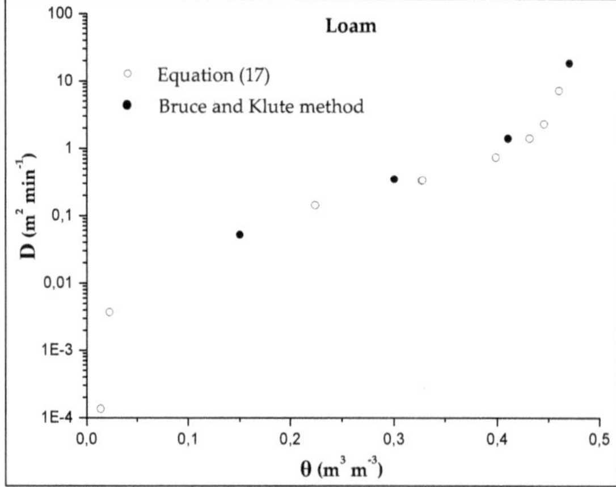

Figure 3. A comparative presentation of the relationship $D(\theta)$ as obtained according to the Bruce and Klute [14] method (Equation (16)) and Equation (17) for the loamy (L) soil.

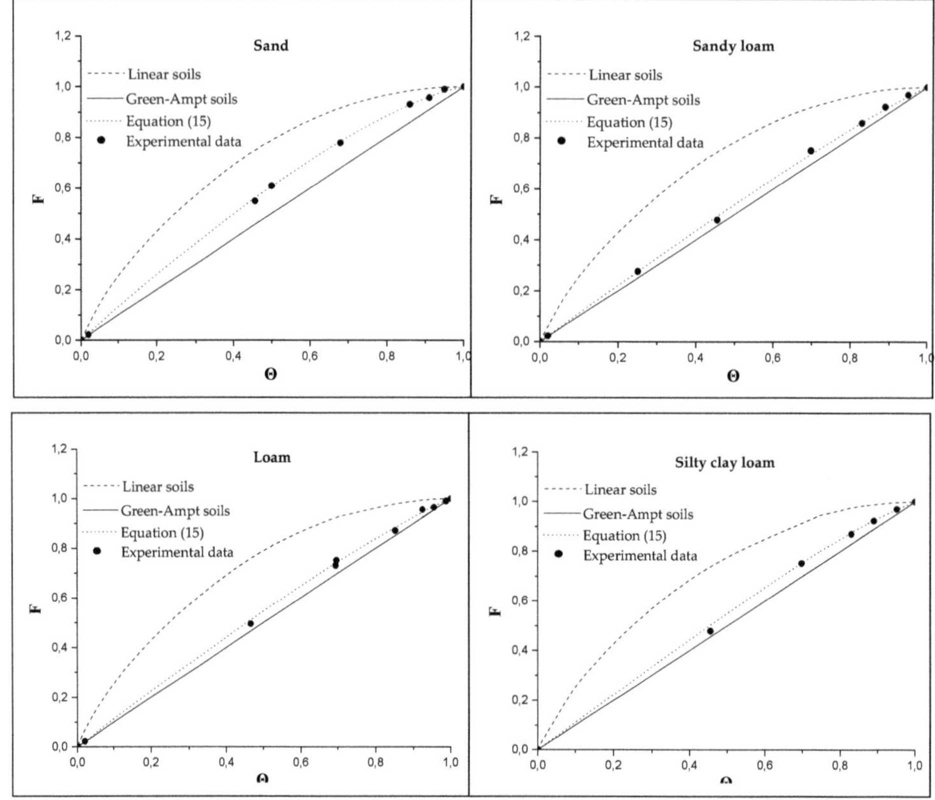

Figure 4. Comparative presentation of the relationship $F(\Theta)$ for the linear soils and Green–Ampt soils, as well as the $F(\Theta)$ calculated according to Equation (15) and the experimental $F(\Theta)$ values described in the text for the four soils examined.

It is worth investigating the ability of the equation $F(\Theta) = 1 - (1 - \Theta)^{p+1}$ to describe the upper and lower theoretical boundaries of F(Θ). As has already mentioned, when $p = 0$ the value of $F = \Theta$—i.e., it converges to the lower limit ($D(\theta)$ function is a delta function). However, the values of F for $p = 2.23$ (resulted from $(p + 1)^{-1} = 0.31$ [8]) differs from the values of F resulting from Equation (13), which is the upper theoretical limit of F(Θ). Specifically, it gives higher values of F(Θ) than the theoretical curve. The fitting of the $F(\Theta) = 1 - (1 - \Theta)^{p+1}$ equation to the theoretical curve (Equation (13)) showed that the equation $F(\Theta) = 1 - (1 - \Theta)^{2.36}$ gives very good results of F(Θ) estimation over the entire range of Θ [7]. Therefore, the variation range of the parameter p is between 0 and 1.36, and the resulting shape parameter value for the linear soils ($(p + 1)^{-1} = 0.423$) is greater than that presented by Clothier et al. [8].

It can be concluded that $F(\Theta) = 1 - (1 - \Theta)^{p+1}$ seems to be appropriate functional form for describing actual flux–saturation curves of general soils, and its parameter p has a physical meaning, i.e., it represents the shape of the soil moisture profile.

Furthermore, one could observe that from the study of Evangelides et al. (Table 2 [6]), where seven different soils are studied, the fitting parameters m (Equation (4) [6]) and $n = p + 1$ (Equation (15)) are strongly linearly related, as $n = p + 1 = -1.483m + 2.4375$. If one uses $m = 4/5$, as Evangelides et al. [6] proposed, then this leads to $\bar{n} = 1.251$ which corresponds to our p value 0.25. In this respect, Equation (4) [6], with $m = 0.8$, becomes equivalent to Equation (15) in the present study, with $p = 0.25$.

5. Conclusions

The mono-parametric equation of the form $(1 - \Theta)^p$ can reliably describe the $\lambda(\Theta)$ relationship after the proper selection of the parameter p for a relatively large range of soils. It is also shown that the analytical expressions of the soil hydraulic diffusivity $D(\theta)$ and soil sorptivity S approach the experimental ones well. Moreover, for the case of the $D(\theta)$ relationship, there is the advantage of obtaining values near saturation, where the classical methodology of Bruce and Klute might be inadequate.

During the laboratory experiment, a new mono-parametric analytical function was used for $F(\Theta)$ evaluation, where the parameter p is estimated from curve-fitting of the experimentally obtained $\lambda(\Theta)$ data to an analytical expression of the form $(1 - \Theta)^p$. Parameter p seems to be strongly related to soil hydraulic properties, and further investigation is needed to find this exact relationship. The analytical $F(\Theta)$ relationship, for all soil types investigated, approaches the experimental ones very well, and lies within the limiting $F(\Theta)$ function for linear and delta function soils. In addition, the upper and lower limit curves of $F(\Theta)$ calculated by the proposed expression were consistent with theoretical curves.

Author Contributions: All authors contributed equally to this work.

Funding: This research received no external funding.

Conflicts of Interest: The authors declare no conflict of interest.

References

1. Philip, J.R. On solving the unsaturated flow equation: 1. The flux-concentration relation. *Soil Sci.* **1973**, *116*, 328–335. [CrossRef]
2. Philip, J.R.; Knight, J.H. On solving the unsaturated flow equation: 3. New quasi-analytical technique. *Soil Sci.* **1974**, *117*, 1–13. [CrossRef]
3. White, I. Measured and approximate flux-concetration lrelations for absorption of water by soils. *Soil Sci. Soc. Am. J.* **1979**, *4*, 1074–1080. [CrossRef]
4. Smiles, D.E.; Knight, J.H.; Perroux, K.M. Absorption of water by soil: The effect of a surface crust. *Soil Sci. Soc. Am. J.* **1982**, *46*, 476–481. [CrossRef]
5. Vauclin, M.; Haverkampt, R. Solutions quasi analytiques de l'equation d'absorption de l'eau par les sols non satires. I. Analyse critique. *Agronomie* **1985**, *5*, 597–606. [CrossRef]
6. Evangelides, C.; Tzimopoulos, C.; Arampatzis, G. Flux-Saturation relationship for unsaturated horizontal infiltration. *Soil Sci.* **2005**, *170*, 671–679. [CrossRef]

7. Ma, D.H.; Zhang, J.B.; Lu, Y.X. Derivation and Validation of a New Soil Pore-Structure-Dependent Flux–Saturation Relationship. *Vadose Zone J.* **2017**, *16*, 1–15. [CrossRef]

8. Clothier, B.E.; Scotter, D.R.; Green, A.E. Diffusivity and One-Dimensional Absorption Experiments. *Soil Sci. Soc. Am. J.* **1983**, *47*, 641–644. [CrossRef]

9. Philip, J.R. General method of exact solutions in concentration-dependent diffusion equation. *Aust. J. Phys.* **1960**, *13*, 1. [CrossRef]

10. McBride, J.F.; Horton, R. An empirical function to describe measured water distributions from horizontal infiltration experiments. *Water Resour. Res.* **1985**, *21*, 1539–1544. [CrossRef]

11. Shao, M.; Horton, R. Integral method for estimating soil hydraulic properties. *Soil Sci. Soc. Am. J.* **1998**, *62*, 585–592. [CrossRef]

12. Ma, D.H.; Wang, Q.; Shao, M.A. Analytical method for estimating soil hydraulic parameters from horizontal absorption. *Soil Sci. Soc. Am. J.* **2009**, *73*, 727–736. [CrossRef]

13. White, I.; Smiles, D.E.; Peroux, K.M. Absorption of water by soil: The constant flux boundary condition. *Soil Sci. Soc. Am. J.* **1979**, *43*, 659–664. [CrossRef]

14. Bruce, R.R.; Klute, A. The measurement of soil water diffusivity. *Soil Sci. Soc. Am. Proc.* **1956**, *20*, 458–462. [CrossRef]

15. Poulovassilis, A. Flow characteristics during infiltration into a horizontal sand column. *Water Resour. Res.* **1977**, *13*, 369–374. [CrossRef]

16. Philip, J.R. Theory of infiltration. *Adv. Hydrosc.* **1969**, *5*, 215–296.

17. Van Genuchten, M.T. A closed-form equation for predicting the hydraulic conductivity of unsaturated soils. *Soil Sci. Soc. Am. J.* **1980**, *44*, 892–898. [CrossRef]

© 2019 by the authors. Licensee MDPI, Basel, Switzerland. This article is an open access article distributed under the terms and conditions of the Creative Commons Attribution (CC BY) license (http://creativecommons.org/licenses/by/4.0/).

Article

Infiltration Characteristics and Spatiotemporal Distribution of Soil Moisture in Layered Soil under Vertical Tube Irrigation

Cheng Wang [1], Dan Bai [1,*], Yibo Li [1], Xinduan Wang [2], Zhen Pei [1] and Zuochao Dong [1]

[1]   State Key Laboratory of Eco-hydraulics in Northwest Arid Region, Xi'an University of Technology, Xi'an 710048, China; chengw8973@163.com (C.W.); liyibo51_teresa@hotmail.com (Y.L.); p18829029376@163.com (Z.P.); dongzuochao2020@163.com (Z.D.)

[2]   School of Science, Xi'an Polytechnic University, Xi'an 710048, China; wxdcdyan@163.com

*   Correspondence: baidan@xaut.edu.cn; Tel.: +86-029-8231-2780

Received: 31 July 2020; Accepted: 25 September 2020; Published: 29 September 2020

Abstract: The limited quantity of irrigation water in Xinjiang has hindered agricultural development in the region and water-saving irrigation technologies are crucial to addressing this water shortage. Vertical tube irrigation, a type of subsurface irrigation, is a new water-efficient technology. In this study, field and laboratory experiments were conducted to analyze (1) the infiltration characteristics and spatiotemporal distribution of moisture in layered soil and (2) the water-saving mechanism of vertical tube irrigation. In the field experiments, we analyzed jujube yield, irrigation water productivity (IWP), and soil moisture in the jujube root zone. In the laboratory irrigation experiments, two soil types (silty and sandy loam) were selected to investigate homogeneous and layered soil, respectively. Cumulative infiltration, wetting body, and soil water moisture distribution were also analyzed. Relative to surface drip irrigation, vertical tube irrigation resulted in slightly lower jujube yields but higher savings in water use (47–68%) and improved IWP. The laboratory experiments demonstrated that layered soil had less cumulative infiltration, a larger ellipsoid wetted body, slower vertical wetting front migration (hindered by layer interface), and faster horizontal wetting front migration than homogenous soil had. The irrigation amount for vertical tube irrigation decreased in layered soil, and water content increased at the layer interface. Vertical tube irrigation in layered soil facilitates the retention of water in the root zone, prevents deep leakage, reduces irrigation amount, and improves the IWP of jujube trees. This study aids the popularization and application of vertical tube irrigation technology.

Keywords: vertical tube irrigation; layered soil; soil moisture distribution; water-saving irrigation

1. Introduction

Water shortages constitute the main factor restricting agricultural development in arid and semiarid areas. Water-efficient irrigation technology is a key means of promoting agricultural development. Subsurface irrigation can effectively save irrigation water by directly transporting water and fertilizer to the rhizosphere through irrigation emitters buried in the soil. Furthermore, it can reduce the irrigation amount and the amount of evaporation, and improve crop yield and water use efficiency [1–6]. Subsurface irrigation technologies include many types, such as subsurface drip irrigation [7–9], porous ceramic irrigation [10], and trace quantity irrigation [11]. Although these technologies are widely used in agriculture, their wide-scale application is constrained by complex manufacturing processes, high working head, and emitter plugging [12,13]. Many experiments and numerical simulations of water flow into soil, the wetting body, soil moisture distribution, and influencing factors of irrigation techniques have been performed [14–17].

Vertical tube irrigation is a new subsurface irrigation technology in which the emitter, a tube with a diameter of 10–20 mm [18], is vertically embedded in the soil. The upper end of the emitter is connected to a water supply pipe, and the lower end of the emitter is open and tightly connected to the soil. Generally, the working head is less than 1 m. Vertical tube irrigation can continually supply water to the plant throughout the irrigation cycle. Bai et al. [18,19] reported the influence of various factors on irrigation discharge, on the spatiotemporal distribution of soil moisture, and on the infiltration process in homogeneous soil. These factors include the working head, vertical tube diameter, physical properties of the soil, and depth of the outlet of the vertical tube emitter.

The total planted area of jujube trees in China accounts for more than 50% of the global total, and China is the primary jujube exporter. Jujube is mainly cultivated in southern Xinjiang, and scientifically designed irrigation is required for high yield. Given the extreme water shortage in southern Xinjiang, we aimed to apply vertical tube irrigation to jujube cultivation to reduce the amount of irrigation water used. The effects of irrigation are closely related to the soil. Furthermore, the complex structure of soil, especially layered soil, affects the distribution of moisture and the growth of crops. Many laboratory and field infiltration experiments have been performed to uncover the rules governing water movement in layered soil [20,21]. The texture and thickness of layered soil affect infiltration characteristics, and the wetting front may stagnate at the layer interface. The wetting front cannot enter the next layer until the soil water moisture of the upper layer is saturated, especially when hydraulic conductivity of the lower layer is greater than that of the upper layer. Hence, layer interfaces may hinder the infiltration process [22,23]. Water movement in layered soil is not a continuous process [24,25]; when the soil matric potential of the lower layer of soil is greater than that of the upper layer, the water enters the lower layer of soil through the layer interface [22].

The water infiltration of vertical tube irrigation in soil is three-dimensional, and it is more complex than simple one-dimensional infiltration. Li et al. [26] used simulations to analyze the soil water movement of subsurface drip irrigation emitters in layered soil, but the problem of water resistance caused by layer interfaces has not been solved. Therefore, we performed field and laboratory infiltration experiments on vertical tube irrigation to study the spatiotemporal distribution of soil moisture in layered soil. Our findings provide a basis for the future design and management of vertical tube irrigation systems in layered soil. Our research aims are: (1) To determine the effects of vertical tube irrigation on jujube yield and IWP of jujube tree; (2) to analyze the spatiotemporal distribution characteristics of soil moisture with vertical tube irrigation in layered soil; and (3) to analyze the water saving mechanism underlying vertical tube irrigation in layered soil.

2. Materials and Methods

Both field and laboratory experiments were performed using subsurface vertical tube irrigation. In the first set of field experiments, we measured the infiltration and soil moisture distribution of the jujube root zone. The second set of experiments was performed in a laboratory.

2.1. Field Experiments

Field experiments of vertical tube irrigation were performed at the jujube experimental base (80°50′ E, 40°29′ N) in Alear, Xinjiang, China (Figure 1). The experimental object was a 6-year-old jujube planted in a plot with a width of 2.5 m and a length of 3.0 m (Figure 2). The soil profile exhibited a clear stratified structure. Soil at the layer depth of 0–50 cm was silty loam, and that at 50–200 cm was sandy loam. Soil bulk density and the initial water content of upper and lower soil layers were measured by drying soil samples in an oven. Soil samples were collected from the field plot and loaded into a steel ring (volume is 100 cm^3, height is 5.0 cm). All soil samples were soaked for 24 h to measure saturated water content. The physical properties of the soil samples are presented in Table 1.

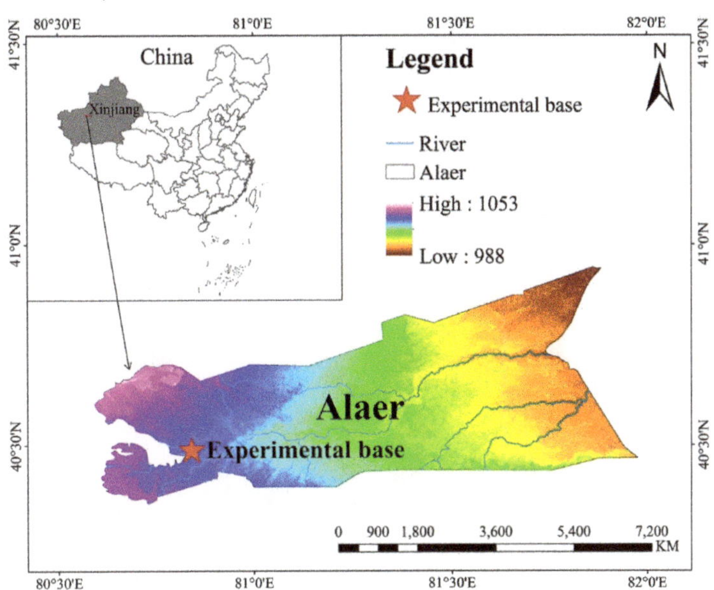

Figure 1. Location of the experimental base.

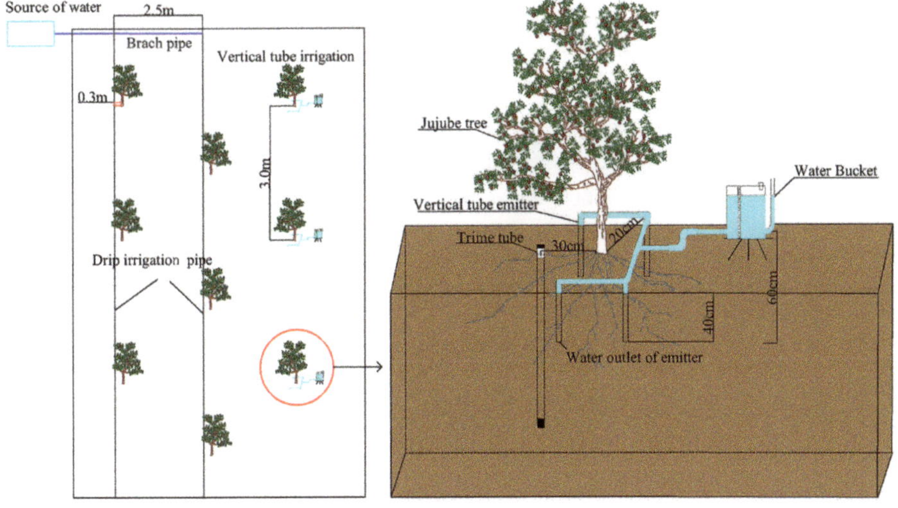

Figure 2. The field experiment setup for vertical tube irrigation and drip irrigation.

The test field was 27 m in length and 20 m in width. Three jujube trees were selected for vertical tube irrigation, and the remaining trees received surface drip irrigation. Spring irrigation was conducted on 15 March 15 2017, with an irrigation amount of 31.03 mm. Spring irrigation is a flood irrigation technique that is usually applied to saline-alkali areas every spring. It can desalinize the soil in root zone and leach the salt accumulated through saline water irrigation. Vertical tube irrigation (11 June–24 July 2017) was performed during the jujube blossom and young fruit period and the developing fruit period. The height difference between the air inlet of the water supply equipment (cuboid water bucket made from the principle of Markov bottle, length × width × height: 29.6 cm × 24.6 cm × 35 cm), and the water outlet of the vertical tube was at 0.6 m (pressure head). The vertical tube emitter (made of polyvinyl chloride) had a length of 450 mm and an inner diameter of 16 mm. The upper end of the vertical tube emitter was connected to a water supply pipe, and the lower end outlets were buried 40 cm below the soil surface. Each tree had four vertical tube emitters (20 cm away from the jujube) to supply water (Figure 2). Each treatment was regarded as one experimental plot (each treatment had one jujube), and each treatment was repeated three times. Surface drip irrigation was applied to the other jujube trees by using one lateral pipe parallel to the row of trees (30 cm from the jujube trees). Drip emitter spacing was 20 cm, and the discharge was 4 L·h^{-1}. Each tree was irrigated three times (on 11 June, 23 June, and 24 July), for 12 h each time. Water meters were installed at the front end of each capillary, and used to record the irrigation amount for drip irrigation.

Table 1. Physical properties of soil layers used in this experiment.

| | Content/% | | | Soil Texture [b] | ρ/g·cm^{-3} | θ_i/cm^3·cm^{-3} | θ_s/cm^3·cm^{-3} |
	Clay <0.002 [a]	Silt 0.002–0.2	Sand 2–0.2				
Field	6.64	47.71	45.64	Silty loam	1.44	0.223	0.446
	3.95	28.5	67.55	Sandy loam	1.47	0.221	0.437
Laboratory	6.78	47.01	46.21	Silty loam	1.44	0.098	0.419
	5.57	30.85	63.58	Sandy loam	1.47	0.047	0.367

[a] Particle diameter. ρ: soil bulk density; θ_i: initial water content; θ_s saturated water content. [b] Classified according to the international system.

To measure the soil moisture distribution, a TRIME tube was installed 1.5 m deep, and 0.3 m away from the tree. The soil profile was measured at 20 cm intervals over the 100-cm-deep soil layer. The soil moisture content was measured at 9:00 a.m. every morning during the whole field experiment. The fruit yield of each jujube was calculated by weight.

2.2. Laboratory Experiments

To further uncover the influence of soil layering on the infiltration and water saving mechanism of vertical tube irrigation when applied to jujube tree, laboratory experiments were performed at the State Key Laboratory of Eco-hydraulics in Northwest Arid Region, Xi'an University of Technology, on November 2018. Experimental soil samples were collected from Xi'an's Loess Plateau. Soil textures were tested using a laser particle size analyzer (MS 2000, Malvern, England, United Kingdom), according to the international soil classification standard (Table 1). The basic physical properties of the experimental soils are displayed in Table 1.

The van Genuchten model [27] was used to generate soil water retention curves:

$$\theta(h) = \begin{cases} \theta_r + \dfrac{\theta_s - \theta_r}{\left(1 + |\alpha h|^n\right)^m} & h < 0 \\ \theta_s & h \geq 0 \end{cases} \tag{1}$$

where, $\theta(h)$ is the soil water content (cm^3 cm^{-3}), h is the pressure head (cm), θ_r is the residual water content (cm^3. cm^{-3}), θ_s is the saturated water content (cm^3 cm^{-3}), m, n, and α are empirically fitted parameters, and $m = 1 - 1/n$.

RETC [28] is a software package for analyzing the relationship between soil water content and matric potential reflections, formulated by van Genuchten in 1980. RETC has the advantages of being able to quantify soil water retention and use hydraulic conductivity functions. Based on the data (θ_s, ρ, and soil particle diameter content) in Table 1, parameters in the van Genuchten model were fitted using RETC and are listed in Table 2.

Table 2. Parameters of van Genuchten model for considered soil.

Soil Texture	$\theta_r/\text{cm}^3\cdot\text{cm}^{-3}$	α/m^{-1}	n	m	$K_s/\text{cm}\cdot\text{min}^{-1}$
Silty loam	0.0357	0.0129	1.4874	0.3277	0.0234
Sandy loam	0.0476	0.0359	3.4067	0.7065	0.5356

θ_r: residual soil water; m, n, and α: empirically fitted parameters, $m = 1 - 1/n$; K_s: hydraulic conductivity properties.

The experimental setup mainly consisted of four parts: the vertical tube emitter, soil box, water supply device, and soil moisture measurement system, as illustrated in Figure 3. The vertical tube emitter (made of polyvinyl chloride) had a length of 330 mm and an inner diameter of 12 mm. The laboratory infiltration experiment was performed in a transparent soil container (made of polymethyl methacrylate), where one-fourth of the container was cylindrical in shape with a height of 50 cm and a bottom radius of 20 cm (net dimensions). The soil container was 0.2 cm thick. The two-side wall plate of the soil container had circular holes with apertures of 1.0 cm (diameter), and the distance between the centers of the holes was 5.0 cm. The upper end of the vertical tube emitter was connected to a Markov bottle (5.0 cm in diameter) by a rubber tube, and the lower end was buried in the soil. Four soil moisture sensor probes (EC-5, Decagon Devices Inc., Pullman, WA, USA) were installed in the soil at different locations near the vertical tube emitter. The intersection of the edge of the one-fourth cylindrical soil container with the surface soil was taken as the origin (0,0). With the surface soil as the reference plane, the positive z direction pointed downward and the positive x direction pointed to the right. The transverse and vertical coordinates of the soil moisture sensor probes of (5, 20), (0, 18), (4, 25), and (3, 27), were denoted as A, B, C, and D, respectively. In the layered soil test, A, B, and C were buried in silty loam, whereas D was buried in sandy loam. The soil water content was measured and recorded at a time interval of 1 min. The experimental setup is detailed in Table 3.

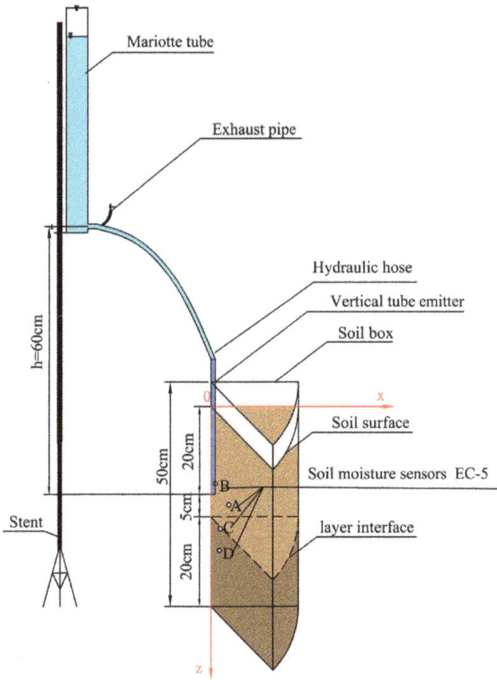

Figure 3. Laboratory experimental setup and instrumentation.

Table 3. Experimental soil stratification.

Depth/cm	Homogeneous Soil	Layered Soil
0–25		Silty loam
25–45	Silty loam	Sandy loam

Infiltration experiments on vertical tube irrigation were performed using homogeneous and layered soil. The specific test arrangements are displayed in Table 3. Soil samples were air dried and sieved through a 2-mm mesh. According to the scheme (Table 3) configuration, soil was compacted into 5 cm layers, which required close contact. Considering the connection between the pipe fittings and the continual supply of water, h_0 was set as 0.6 m (the pressure head during the operation of the vertical tube irrigation emitter). The vertical tube was buried 20 cm below the surface (Figure 3).

The total infiltration time was 8 h for each treatment. During the first 10 min, the recorded times were 0, 1, 3, 5, and 10 min. Between 10 and 60 min, the time interval was 10 min for every record. Between 1 and 3 h, the time interval was 20 min for every record, and between 3 and 8 h, the last interval was 1 h. At different infiltration times, the water level in the Markov bottle was observed, and the wetting front was recorded. A cover film was applied to the soil surface to prevent evaporation. After infiltration, soil water content was measured using an oven drying method with soil samples taken from various depths of the soil (depths of 13, 16, 18, 20, 25, 27, and 30 cm). Infiltration tests of homogeneous soil and layered soil were repeated in groups of three, and the test values were the mean values of three tests.

2.3. Irrigation Water Productivity

Irrigation water productivity (IWP) [29] is defined as the ratio of irrigation water amount to crop yield. The IWP intuitively captures the effect on crop yields from the amount of irrigation water consumed, and it is thus a widely used indicator in agriculture for evaluating the effect of irrigation. IWP was calculated using the following equation:

$$IWP = 100\frac{Y}{I} \tag{2}$$

where *IWP* is expressed in kg·m^{-3}, *Y* is the jujube yield (Mg ha^{-1}), and *I* is the irrigation amount (mm).

3. Results

3.1. Field Experiment Results

3.1.1. Jujube Tree Yield and Water Productivity

Jujube trees receiving vertical tube irrigation were irrigated continually for 44 days during the entire growth period. Surface drip irrigation was performed three times (according to the local irrigation system), for 12 h each time. Data on jujube yield (*Y*), irrigation water amount (*I*), and IWP for the experimental groups are presented in Table 4.

The two irrigation technologies significantly differed only with respect to amount of irrigation water ($p < 0.05$) and not in terms of yield or IWP. However, the IWP of vertical tube irrigation was 1.6 times that of surface drip irrigation. This may be due to the small discharge but continuous irrigation provided by vertical tube irrigation, in contrast to the cycle large-scale centralized that is characteristic of surface drip irrigation. The water from vertical tube irrigation can be stored in the buckets, whereas the water form surface drip irrigation falls into the soil and is thus more likely to be reduced by evaporation.

Second, vertical tube irrigation is type of subsurface irrigation, and the water is directly transported to the root layer; this improves the use of water by jujube trees. This was why vertical tube irrigation saved water while not compromising jujube yield, thereby improving IWP.

Table 4. Jujube trees yield, irrigation water amount, and irrigation water productivity (IWP) under different treatments in the experimental years.

Irrigation Method	Experiments Number	I/mm	Y/ Mg·ha^{-1}	IWP/kg·m^{-3}
Vertical tube irrigation	1	154	15.80	6.16
	2	254	14.36	3.39
	3	214	14.87	4.17
	*	$207 \pm 41b$	15.01 ± 0.60	$4.57 \pm 0.12a$
Surface drip irrigation	1	489	14.70	2.76
	2	483	15.10	3.02
	3	495	15.47	2.69
	*	$489 \pm 5a$	15.09 ± 0.31	$2.82 \pm 0.14b$

*: Data are in terms of mean ± SD standard deviation of three replicate samples; I: irrigation water amount; Y: jujube yield; IWP: irrigation water productivity.

3.1.2. Soil Moisture Distribution in the Field

Beginning from 24 h after the first irrigation, soil moisture content was measured once per day (at 9:00 a.m.) until October 3 for a total of 115 days. The distribution of soil water content is illustrated in Figure 4. For vertical tube irrigation, soil moisture was >26% at 0–50 cm, and the moisture content at 50–100 cm gradually decreased from 24% (Figure 4a). The outlet of the vertical tube emitter was buried 40 cm below the soil surface, and the water gradually diffused outward. As the soil water moisture of the outlet increases, the discharge of the vertical tube emitter decreases, and the water diffusion rate decreases. In particular, when water reaches the layer interface, the water potential of the upper soil was greater than the lower soil. Therefore, this may hinder the downward migration of water. Furthermore, surface soil moisture gradually evaporates, causing the moisture to migrate upwards, which explains why the soil moisture at the 0–50 cm root layer of silty loam is higher. The water-holding capacity of the upper soil is greater than the water-holding capacity of the lower soil, and less irrigation water enters the lower soil, which makes deep leakage unlikely to occur.

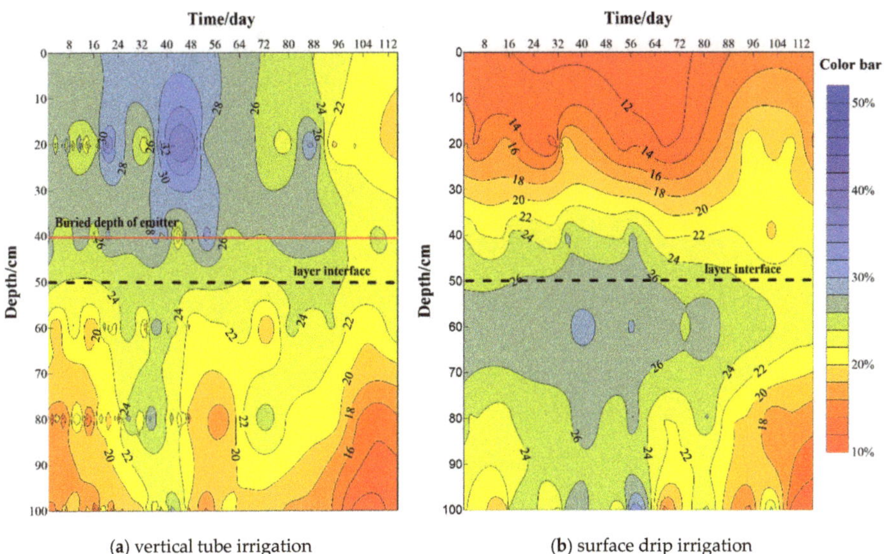

(a) vertical tube irrigation   (b) surface drip irrigation

Figure 4. Vertical distribution of soil moisture content.

Because surface soil moisture is gradually reduced by evaporation, the upper layer (0–50 cm depth from the surface) of the soil is relatively dry. Because of the poor water-holding capacity of sand, under the action of gravity, water can easily move downwards in the lower sandy soil to go even deeper. Most of the jujube roots were concentrated at the surface from 0–50 cm [30], whereas few were distributed in deeper soil. The soil moisture in the main jujube root distribution area can be maintained at a constant range through vertical tube irrigation. However, because all the water was irrigated at every single period during the surface drip irrigation, the large amount of irrigation water could move through the soil interface layer into deeper soil after irrigation, which was not conducive to the use of water by the main roots layer.

3.2. Laboratory Experiment Results

3.2.1. Cumulative Infiltration

Figure 5 illustrates the evolution of the cumulative infiltration for vertical tube irrigation. The infiltration process in layered soil can be divided into two stages [31]: (1) the infiltration of water above the layer interface, and (2) the infiltration of water through the layer interface to the lower soil. In the first stage, the infiltration process in homogeneous soil lasted for approximately 1 h, and the wetting front moved to the position of the layer interface. Before the infiltration front reached the layer interface, the cumulative infiltration of layered soil and homogeneous soil were similar. In the second stage (at 1–8 h), the cumulative infiltration was lower in the layered soil than in the homogeneous soil. The cumulative infiltration amount in the homogeneous and layered soils were 198.63 L and 166.64 L, respectively. The cumulative infiltration in layered soil was 84% of that in homogeneous soil. The cumulative infiltration decreased when vertical tube irrigation in layered soil was used, a finding consistent with that of a previous study [32].

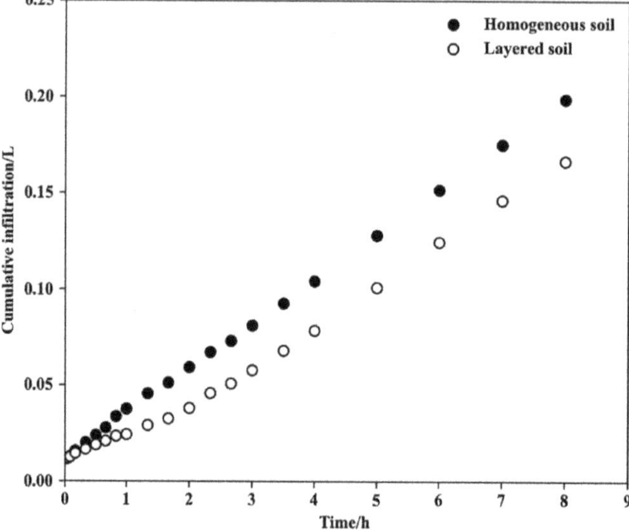

Figure 5. Evolution of cumulative infiltration.

3.2.2. Wetted Body Changes with Time

Figure 6 illustrates the variations in the wetted body over time. The vertical and transverse coordinates of the vertical tube emitter outlet were (0, 20), and 12 infiltration times (1, 5, 20, 40, 60, 120, 180, 240, 300, 360, 420, and 480 min) were selected to draw the boundary line of the wetted body. Figure 6a,b present variations in the wetted body over time in homogeneous and layered soil, respectively. The wetted body spread from the vertical tube emitter outlet, and the spread rate gradually decreased with infiltration time. Similar to that in homogeneous soil, the volume of the upper hemisphere of the wetting body was greater than that of the lower hemisphere. Because the lower layer was sandy soil with poor water-holding capacity and rapid water diffusion, the wetted body in layered soil was slightly larger than that in homogeneous soil.

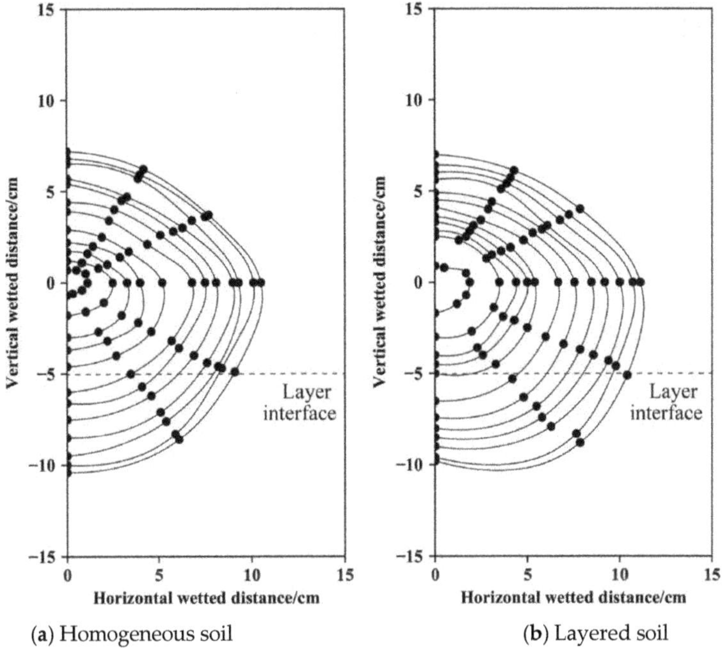

(a) Homogeneous soil   (b) Layered soil

Figure 6. Variations in wetted body with time. The boundary line of the wet body corresponds to 1, 5, 20, 40, 60, 120, 180, 240, 300, 360, 420, and 480 min from the inside to the outside.

Because the water-holding capacity of the upper soil was greater than that of the lower soil, the water potential difference at the layer interface affected the downward migration of moisture; under the influence of the soil matric potential, the moisture diffused horizontally. As soil moisture content increased, upper soil suction decreased gradually until it was equal to or less than the air intake suction of the lower sandy loam, and moisture moved downward through the layer interface. Because sandy soil has high water conductivity, any wetted body in sandy soil tends to diffuse rapidly. Therefore, the lower hemisphere body in layered soil was larger than that in homogeneous soil.

3.2.3. Variation in Transport Distance of Wetting Front

Figure 7 illustrates the variation in maximum wetting front transport distance over time in the upward, horizontal, and downward directions. After 8 h of infiltration, the upward migration distance was the smallest, and the difference between the horizontal and downward wetting front distances was nonsignificant. In the upward and horizontal direction, the maximum wetting front transport distance was determined by the soil texture. The transport distance of the downward wetting front in layered soil was not significantly different from that in homogeneous soil because the upper soil had a large water-holding capacity. Although less moisture entered the lower layer, the wetting front diffused quickly because of the high hydraulic conductivity of the sandy soil.

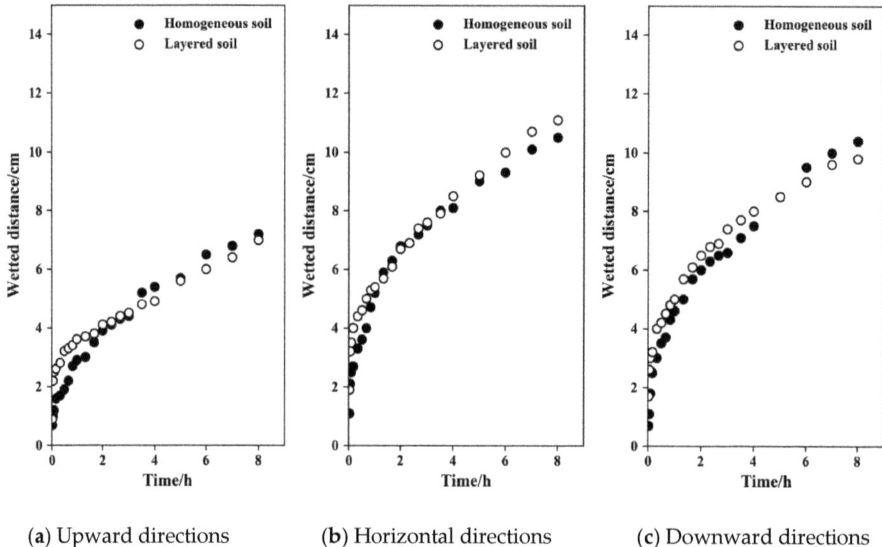

(**a**) Upward directions   (**b**) Horizontal directions   (**c**) Downward directions

Figure 7. Evolution of maximum wetting front movement distance.

3.2.4. Effect of Layer Interface on Water Infiltration

To explore the influence of the layer interface on water infiltration in vertical tube irrigation, movements of the wetting front above and below the layer interface were analyzed. Figure 8 presents the evolution of wetting front migration distance in layered soil, and Figure 9 presents the evolution of wetting front migration rate over time.

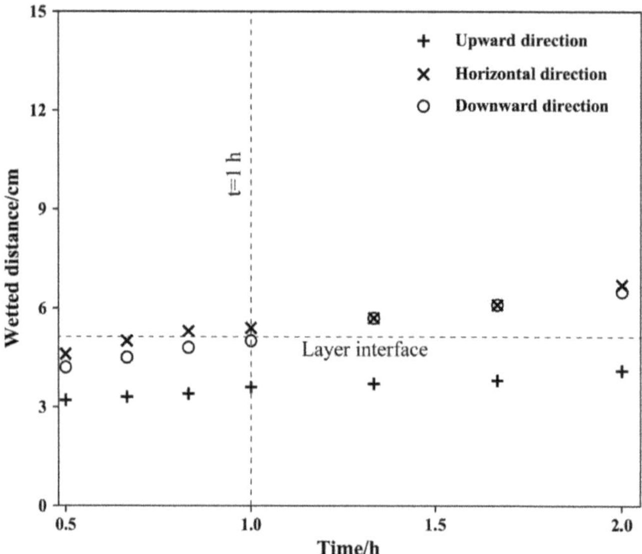

Figure 8. Evolution of wetting-front migration distance in layered soil.

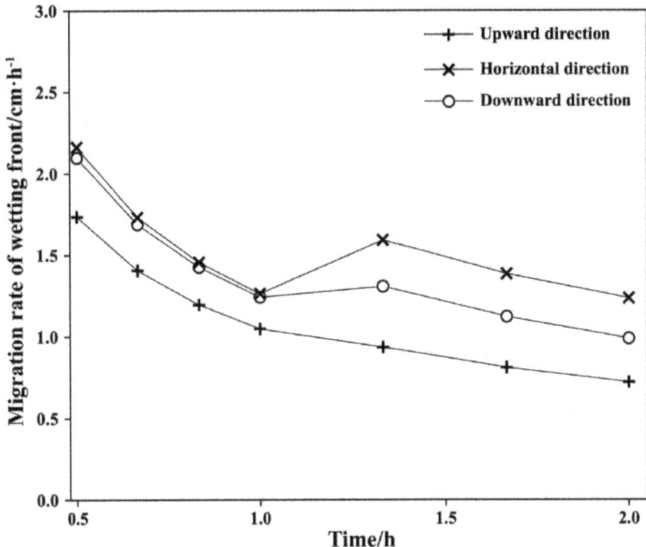

Figure 9. Migration of wetting front in layered soil.

In Figure 8, when the infiltration time was less than 1 h, the upward wetting front migration distance was the smallest. The horizontal wetting front migration distance was slightly smaller than that of the downward wetting front, and the wetting front migration rate decreased. When the infiltration time was 1 h, the downward wetting-front migration distance reached 5.0 cm (i.e., water migrated to the layer interface). The downward wetting-front movement was blocked, and the wetting-front movement rate reached the first trough (Figure 9).

As shown in Figure 9, when the infiltration time was greater than 1 h, the downward and horizontal wetting front migration rates increased, and the horizontal wetting front migration rate increased faster than that of the downward wetting front. After 80 min of infiltration, the horizontal and vertical wet front movement rates began to decrease simultaneously. During the infiltration process, the vertical upward wetting-front movement rate gradually decreased over time.

Water transport in layered soil was affected by the layer interface. The wetting front arrives at the layer interface—the infiltration rate is believed to decrease at the wetting front arrival time to the interface of the two layers—because the attraction is reduced. Water infiltration in homogeneous soil was affected by soil and gravity during the infiltration stage before reaching the layer interface. After reaching and passing through the layer interface, water could not move downward and accumulated in the upper layer. This promoted horizontal water movement and increased horizontal wetting front speed.

Layered soil infiltration, three-dimensional vertical pipe irrigation, and water were considerably affected by the layer interface. The layer interface hinders downward migration of moisture but promotes water transport in other directions, particularly horizontally. Therefore, the wetting front rate fluctuated at the layer interface. In addition, the water migration rate was affected by the high hydraulic conductivity of sand. At larger permeabilities of the lower soil layer, the advancing speed of the wetting front increases [20].

3.2.5. Variation in Soil Water Content

Figure 10a,b illustrate variations in soil moisture content over time during homogeneous and layered soil infiltration. During homogeneous soil infiltration, soil moisture over time gradually increased. At the end of infiltration (8 h), the soil moisture content reached 32.6% at point C and 27.8%

at point A. The moisture content at points B and D (particularly point B) lagged because of the effect of gravity, which promoted the movement of water. In layered soil, moisture content at points A, B, and C gradually increased over the first hour, whereas that above the silty loam layer interfaces increased relatively rapidly. After infiltration, the largest change was at point C, which reached 37.6% moisture content. Points A and B had 37.2% and 31.6%, respectively. Point D in the sand below the layer interface reached 11.4% because the layer interface hindered moisture diffusion into the sand below. In this layered soil in which the upper layer is silty soil and the lower layer is sandy soil, the upper layer of the soil aided the retention for water, improving the moisture content above the layer interface. Suitable soil water content in root zones is beneficial for crop growth [1]. Vertical tube irrigation in layered soil can effectively reduce the amount of water infiltration and increase water content in the upper soil and maintain the soil water moisture within a constant range.

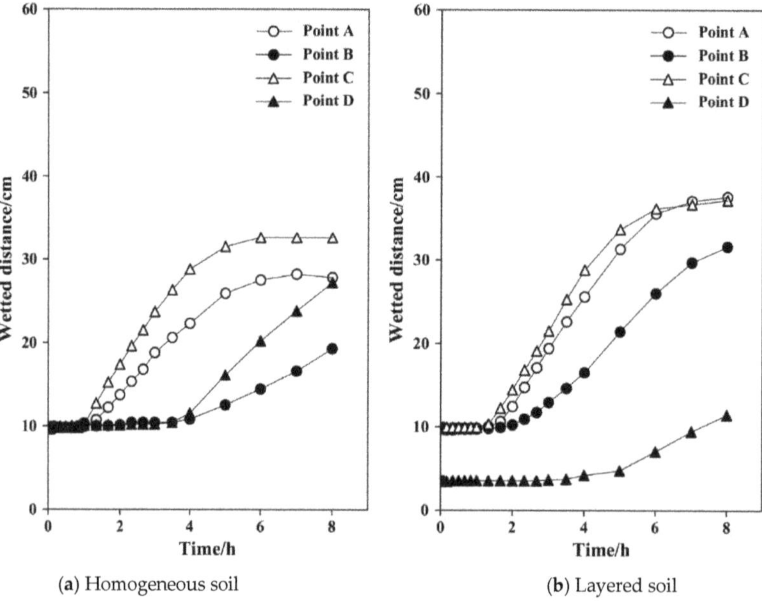

(**a**) Homogeneous soil   (**b**) Layered soil

Figure 10. Migration of soil moisture content over time monitored using soil sensors.

As presented in Figure 11, after infiltration (8 h), the soil moisture content reached a maximum (close to saturation) at the vertical tube irrigation outlet (depth = 20 cm). The soil moisture contents of homogeneous and layered soil were 39.13% and 40.08%, respectively. Ponding infiltration is one-dimensional, and after the upper soil is saturated, water begins to migrate vertically and exhibits a "finger flow" [33] phenomenon in the soil. However, in the process of vertical tube infiltration, the water content at the outlet was the highest, and the layer interface changes the shape of the wetted body instead of generating preferential flow, as is the case in one-dimensional vertical infiltration.

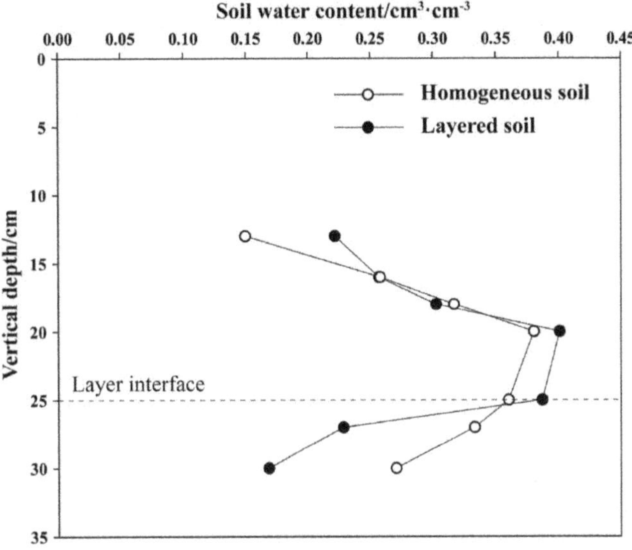

Figure 11. Soil moisture content by depth.

4. Discussion

4.1. Effects of Vertical Tube Irrigation on Jujube Yield and IWP

Irrigation amount and jujube yield were compared between surface drip and vertical tube irrigation. Vertical tube irrigation yielded obvious savings in water, at an irrigation amount of 207 ± 41 mm compared with 489 ± 5 mm for drip irrigation. After surface irrigation, the soil moisture content increased rapidly after irrigation and then gradually decreased because of the redistribution and evaporation of water in the soil [8]. Under high temperature, a large amount of soil moisture evaporates. Thus, in soils with a surface depth of 0–50 cm, the soil water content of surface drip irrigation gradually increased, but only to 12–20%. Vertical tube irrigation is a type of subsurface irrigation. The burial depth of the vertical tube water outlet was 40 cm, and the surface evaporation was small. Therefore, soil water content within 10–40 cm was maintained between 26% and 35% in the jujube field experiment with vertical tube irrigation.

Furthermore, the soil profile exhibited an obvious stratified structure, and the hydraulic conductivity was lower in the upper layer than in the lower layer. For surface drip irrigation technology, water gradually moved downward from the surface, and when water passes through the layer interface, it soon enters the sandy loam and spreads. Rawls et al. [34] found that sandy soil has high hydraulic conductivity and much less holding capacity under conditions of high soil water content. The maximum water content for surface drip irrigation was observed 50–80 cm below the surface. The soil from 50–200 cm was sandy loam, and the water was easily transported downward. However, because the main roots of jujube are distributed in the 0–50 cm soil layer below the surface, the plant has difficulty absorbing water in this part of the soil, resulting in low IWP of surface drip irrigation. Similar to porous ceramic irrigation [10], we noted a feedback regulation effect between the discharge of vertical tube irrigation and soil water content. When the soil water content was high, the discharge of vertical tube irrigation decreased; furthermore, because the soil profile exhibited an obvious stratified structure, the layer interface reduced the downward migration of water, and kept the water in the main root layer area. The IWP of vertical tube irrigation was 1.6 times higher than that of surface drip irrigation. The results demonstrated that this limited water supply could reduce jujube water consumption and that a greater abundance of soil water can support jujube growth, as also shown by Ma et al. [35].

4.2. Infiltration Characteristics in Layered Soil with Vertical Tube in Laboratory

In a laboratory, the infiltration law of vertical tube irrigation in layered soil and the spatiotemporal distributions of soil moisture were analyzed. The moisture content of wetted layered soil below the layer interface was less than that of homogenous soil in the same position. Yang et al. [36] demonstrated that the finer layer with low saturated permeability limited the flux into the coarser layer. The moisture in layered soil passed through the layer interface and was affected by the difference in soil water potential, which caused water to accumulate in the upper soil and gradually increase the soil water content. When the influent suction in the sandy soil was greater than the soil water potential in the upper soil, under the action of gravity potential, the water began to move downward, which reduced the infiltration amount and the irrigation amount.

In contrast to that of underground drip irrigation, the discharge of vertical tube irrigation was mainly affected by soil water potential. Reducing the irrigation amount can effectively maintain soil moisture in the main root zones. Improving the soil water content in the main root zones can increase root-to-water use efficiency [37]. However, the burial depth of the water outlet of the vertical tube emitter must be determined according to the observed soil profile structure, in doing so, users can fully leverage the advantages of this irrigation technology.

5. Conclusions

This study performed field and laboratory experiments on vertical tube irrigation. When jujube trees were subject to vertical tube irrigation and surface drip irrigation, vertical tube irrigation had the (slightly) lower jujube yield but better (i.e., smaller) irrigation amount, at approximately 42.3% of the irrigation amount of surface drip irrigation. The IWP of vertical tube irrigation was 1.6 times that of surface drip irrigation. The water content for vertical tube irrigation was mainly concentrated in the root distribution area of the soil (0–50 cm deep), which improved the root system's water utilization rate. Vertical tube irrigation makes good use of the water barrier effect of the layer interface. The layer interface in vertical tube irrigation prevented some downward movement of water. Moisture was effectively prevented from spreading to the sand, and the moisture content of the upper soil increased. It promoted the horizontal diffusion of the wetting front, making the soil moisture content of the main root layer relatively constant and reducing evaporation and deep leakage from the surface.

Vertical tube irrigation is a continuous irrigation method with a small amount of irrigation, but it can maintain a suitable soil moisture content in the root layer of the crop and create a favorable water environment for crop growth. The application of vertical tube irrigation technology in jujube cultivation can economize the use of irrigation water, which can greatly aid jujube cultivation in arid Xinjiang. We conducted laboratory analyses of the water-saving principle underlying vertical tube irrigation under layered soil conditions. These results aid the study of the hydraulic characteristics and soil water distributions that are associated with irrigation systems of the subsurface-drip and porous-ceramic types. However, this paper only discusses a layered method. We intend to analyze various combinations of layered soil and the infiltration characteristics of vertical tube irrigation technology in different soil structures to provide a theoretical basis for the practical application of vertical tube irrigation.

Author Contributions: Data curation, C.W., Z.D. and Z.P.; writing—original draft preparation, C.W.; writing—review and editing, C.W., D.B., X.W. and Y.L.; funding acquisition and supervision, D.B. All authors have read and agreed to the published version of the manuscript.

Funding: This work was supported by the Projects of the National Natural Science Foundation of China, grant numbers 41571222,51909208.

Conflicts of Interest: The authors declare no conflict of interest.

References

1. Patel, N.; Rajput, T.B.S. Effect of drip tape placement depth and irrigation level on yield of potato. *Agric. Water Manag.* **2007**, *88*, 209–223. [CrossRef]
2. Patel, N.; Rajput, T.B.S. Dynamics and modeling of soil water under subsurface drip irrigated onion. *Agric. Water Manag.* **2008**, *95*, 1335–1349. [CrossRef]
3. Abuarab, M.; Mostafa, E.; Ibrahim, M. Effect of air injection under subsurface drip irrigation on yield and water use efficiency of corn in a sandy clay loam soil. *J. Adv. Res.* **2013**, *4*, 493–499. [CrossRef] [PubMed]
4. Mo, Y.; Li, G.; Wang, D. A sowing method for subsurface drip irrigation that increases the emergence rate, yield, and water use efficiency in spring corn. *Agric. Water Manag.* **2017**, *179*, 288–295. [CrossRef]
5. Coltro, L.; Marton, L.F.M.; Pilecco, F.P.; Pilecco, A.C.; Mattei, L.F. Environmental profile of rice production in Southern Brazil: A comparison between irrigated and subsurface drip irrigated cropping systems. *J. Clean. Prod.* **2017**, *153*, 491–505. [CrossRef]
6. Pendergast, L.; Bhattarai, S.P.; Midmore, D.J. Evaluation of aerated subsurface drip irrigation on yield, dry weight partitioning and water use efficiency of a broad-acre chickpea (*Cicer arietinum*, L.) in a vertosol. *Agric. Water Manag.* **2019**, *217*, 38–46. [CrossRef]
7. Li, J.S.; Li, Y.F.; Zhang, H. Tomato Yield and Quality and Emitter Clogging as Affected by Chlorination Schemes of Drip Irrigation Systems Applying Sewage Effluent. *J. Integr. Agric.* **2012**, *11*, 1744–1754. [CrossRef]
8. Martínez, J.; Reca, J. Water use efficiency of surface drip irrigation versus an alternative subsurface drip irrigation method. *J. Irrig. Drain. Eng.* **2014**, *140*, 1–9. [CrossRef]
9. Jacques, D.; Fox, G.; White, P. Farm level economic analysis of subsurface drip irrigation in Ontario corn production. *Agric. Water Manag.* **2018**, *203*, 333–343. [CrossRef]
10. Cai, Y.; Wu, P.; Zhang, L.; Zhu, D.; Chen, J.; Wu, S.J.; Zhao, X. Simulation of soil water movement under subsurface irrigation with porous ceramic emitter. *Agric. Water Manag.* **2017**, *192*, 244–256. [CrossRef]
11. Zhu, J.; Jin, J.; Yang, C. Effects of trace quantity irrigation on yield, dry matter portioning and water use efficiency of spherical fennel grown in greenhouse. *Paiguan Jixie Gongcheng Xuebao/J. Drain. Irrig. Mach. Eng.* **2014**, *32*, 338–342.
12. Song, P.; Li, Y.; Zhou, B.; Zhou, C.; Zhang, Z.; Li, J. Controlling mechanism of chlorination on emitter bio-clogging for drip irrigation using reclaimed water. *Agric. Water Manag.* **2017**, *184*, 36–45. [CrossRef]
13. Zhou, H.; Li, Y.; Wang, Y.; Zhou, B.; Bhattarai, R. Composite fouling of drip emitters applying surface water with high sand concentration: Dynamic variation and formation mechanism. *Agric. Water Manag.* **2019**, *215*, 25–43. [CrossRef]
14. Das Gupta, A.; Babel, M.S.; Ashrafi, S. Effect of soil texture on the emission characteristics of porous clay pipe for subsurface irrigation. *Irrig. Sci.* **2009**, *27*, 201–208. [CrossRef]
15. Kandelous, M.M.; Šimůnek, J. Comparison of numerical, analytical, and empirical models to estimate wetting patterns for surface and subsurface drip irrigation. *Irrig. Sci.* **2010**, *28*, 435–444. [CrossRef]
16. Yao, W.W.; Ma, X.Y.; Li, J.; Parkes, M. Simulation of point source wetting pattern of subsurface drip irrigation. *Irrig. Sci.* **2011**, *29*, 331–339. [CrossRef]
17. Ren, C.; Zhao, Y.; Wang, J.; Bai, D.; Zhao, X.; Tian, J. Lateral hydraulic performance of subsurface drip irrigation based on spatial variability of soil: Simulation. *Agric. Water Manag.* **2017**, *193*, 232–239. [CrossRef]
18. Bai, D.; He, J.; Guo, L.; Wang, L. Infiltration characteristics of vertical tube subirrigation as affected by various factors. *Trans. Chin. Soc. Agric. Eng.* **2016**, *32*, 101–105.
19. Bai, D.; Sun, S.; Ren, P.; Xu, X. Temporal and spatial variation of wetting volume under sub-irrigation with vertical emitter. *Trans. Chin. Soc. Agric. Eng.* **2018**, *34*, 107–113.
20. Mohammadzadeh-Habili, J.; Heidarpour, M. Application of the Green-Ampt model for infiltration into layered soils. *J. Hydrol.* **2015**, *527*, 824–832. [CrossRef]
21. Soulis, K.X.; Elmaloglou, S. Optimum soil water content sensors placement for surface drip irrigation scheduling in layered soils. *Comput. Electron. Agric.* **2018**, *152*, 1–8. [CrossRef]
22. Cho, K.W.; Song, K.G.; Cho, J.W.; Kim, T.G.; Ahn, K.H. Removal of nitrogen by a layered soil infiltration system during intermittent storm events. *Chemosphere* **2009**, *76*, 690–696. [CrossRef] [PubMed]
23. Cui, G.; Zhu, J. Prediction of unsaturated flow and water backfill during infiltration in layered soils. *J. Hydrol.* **2018**, *557*, 509–521. [CrossRef]

24. Ma, Y.; Feng, S.; Su, D.; Gao, G.; Huo, Z. Modeling water infiltration in a large layered soil column with a modified Green-Ampt model and HYDRUS-1D. *Comput. Electron. Agric.* **2010**, *71*, 40–47. [CrossRef]

25. Deng, P.; Zhu, J. Analysis of effective Green-Ampt hydraulic parameters for vertically layered soils. *J. Hydrol.* **2016**, *538*, 705–712. [CrossRef]

26. Li, J.S.; Ji, H.Y.; Bei, L.I.; Liu, Y.C. Wetting Patterns and Nitrate Distributions in Layered-Textural Soils Under Drip Irrigation. *Agric. Sci. China* **2007**, *6*, 970–980. [CrossRef]

27. van Genuchten, M.T. A closed form equation for predicting the hydraulic conductivityof unsaturated soils. *Soil Sci. Soc. Am. J.* **1980**, *44*, 892–898. [CrossRef]

28. van Genuchten, M.; Leij, F.J.; Yates, S.R. *The RETC Code for Quantifying the Hydraulic Functions Unsaturated Soils*; EPA/600/2-91/065; R.S. Kerr Environmental Research Laboratory, U.S. Environmental Protection Agency: Okla, LA, USA, 1991.

29. Monaco, F.; Sali, G. How water amounts and management options drive Irrigation Water Productivity of rice. A multivariate analysis based on field experiment data. *Agric. Water Manag.* **2018**, *195*, 47–57. [CrossRef]

30. Dai, Z.; Fei, L.; Huang, D.; Zeng, J.; Chen, L.; Cai, Y. Coupling effects of irrigation and nitrogen levels on yield, water and nitrogen use efficiency of surge-root irrigated jujube in a semiarid region. *Agric. Water Manag.* **2019**, *213*, 146–154. [CrossRef]

31. Li, Y.; Ren, X.; Hill, R.; Malone, R.; Zhao, Y. Characteristics of Water Infiltration in Layered Water-Repellent Soils. *Pedosphere* **2018**, *28*, 775–792. [CrossRef]

32. Huang, M.; Barbour, S.L.; Elshorbagy, A.; Zettl, J.; Si, B.C. Effects of Variably Layered Coarse Textured Soils on Plant Available Water and Forest Productivity. *Procedia Environ. Sci.* **2013**, *19*, 148–157. [CrossRef]

33. Allaire, S.E.; Dadfar, H.; Denault, J.T.; van Bochove, E.; Charles, A.; Thériault, G. Development of a method for estimating the likelihood of finger flow and lateral flow in canadian agricultural landscapes. *J. Hydrol.* **2011**, *403*, 261–277. [CrossRef]

34. Rawls, W.J.; Gish, T.J.; Brakensiek, D.L. Estimating soil water retention from soil physical properties and characteristics. *Adv. Soil Sci.* **1991**, *16*, 213234.

35. Ma, L.; Wang, X.; Gao, Z.; Wang, Y.; Nie, Z.; Liu, X. Canopy pruning as a strategy for saving water in a dry land jujube plantation in a loess hilly region of China. *Agric. Water Manag.* **2019**, *216*, 436–443. [CrossRef]

36. Yang, H.; Rahardjo, H.; Leong, E.C. Behavior of unsaturated layered soil columns during infiltration. *J. Hydrol. Eng.* **2006**, *11*, 329–337. [CrossRef]

37. Wang, J.; Niu, W.; Li, Y.; Lv, W. Subsurface drip irrigation enhances soil nitrogen and phosphorus metabolism in tomato root zones and promotes tomato growth. *Appl. Soil Ecol.* **2018**, *124*, 240–251. [CrossRef]

© 2020 by the authors. Licensee MDPI, Basel, Switzerland. This article is an open access article distributed under the terms and conditions of the Creative Commons Attribution (CC BY) license (http://creativecommons.org/licenses/by/4.0/).

Article

Analytic Representation of the Optimal Flow for Gravity Irrigation

Carlos Fuentes [1] and Carlos Chávez [2,*]

[1] Mexican Institute of Water Technology, Paseo Cuauhnáhuac Núm. 8532, Jiutepec 62550, Morelos, Mexico; cbfuentesr@gmail.com

[2] Water Research Center, Department of Irrigation and Drainage Engineering, Autonomous University of Queretaro, Cerro de las Campanas SN, Col. Las Campanas, Queretaro 76010, Mexico

* Correspondence: chagcarlos@uaq.mx; Tel.: +52-442-192-1200 (ext. 6036)

Received: 7 September 2020; Accepted: 26 September 2020; Published: 27 September 2020

Abstract: The aim of this study is the deduction of an analytic representation of the optimal irrigation flow depending on the border length, hydrodynamic properties, and soil moisture constants, with high values of the coefficient of uniformity. In order not to be limited to the simplified models, the linear relationship of the numerical simulation with the hydrodynamic model, formed by the coupled equations of Barré de Saint-Venant and Richards, was established. Sample records for 10 soil types of contrasting texture were used and were applied to three water depths. On the other hand, the analytical representation of the linear relationship using the Parlange theory of infiltration proposed for integrating the differential equation of one-dimensional vertical infiltration was established. The obtained formula for calculating the optimal unitary discharge is a function of the border strip length, the net depth, the characteristic infiltration parameters (capillary forces, sorptivity, and gravitational forces), the saturated hydraulic conductivity, and a shape parameter of the hydrodynamic characteristics. The good accordance between the numerical and analytical results allows us to recommend the formula for the design of gravity irrigation.

Keywords: Saint-Venant equations; Richards' equation; Parlange equations; optimal irrigation flow; soil parameters; analytical representation

---

1. Introduction

Gravity irrigation is the water supply at the head of a channel or inclined ditch built on a plot, as a border or a furrow, in order to take advantage of the gravitational field to provide the necessary amount of water for optimal development of cultivated plants. In continuous gravity irrigation, three phases are distinguished in the surface water movement [1,2]. The first begins when water flow is provided on the dry border until the water wave reaches the end part of the same; it is known as the advance phase. The second starts from the arrival of the wave at the end of the border until the water supply is cut off, known as the storage phase. Finally, the third phase, known as the recession, is composed of two sub-phases, one is the vertical recession starting from cutting off the water supply until the depth at the head of the border disappears, and the other is the horizontal recession starting from the disappearance of the depth at the head and ends when the depth at the end of the border disappears.

According to the principles used in the modeling, studies reported in literature can be grouped in the context of four approaches [2]: (1) The modeling of surface and underground movements is addressed in a full empirical way [3]; (2) the surface movement is modeled with the Barré de Saint-Venant equations and their simplifications, and the underground movement with empirical equations as those of Kostiakov and Mezencev [4–11]; (3) surface movement is modeled by Saint-Venant simplified equations (kinematic wave, diffusion wave or null inertia and hydrological model) and in the modeling of the underground movement there is the possibility of using rational equations [12–23]; (4) the surface movement is modeled with Barré de Saint-Venant equations and the underground movement with Richards' equation [24] in [2,25–28].

In the study to model the border irrigation developed by Schmitz et al. [29], the complete equations of Barre de Saint-Venant are solved with the method of characteristics, and for the underground movement, the analytical solution for infiltration obtained by Parlange et al. is used [30]. Other studies use the Green and Ampt infiltration equation [31], which is a special case of Richard's two-dimensional equation [32], and other group coupled Saint-Venant and Richards equations in border irrigation. The first ones are resolved with a Lagrangian method and the second one with the finite element method [25–28]. In Saucedo et al. [27], using the full hydrodynamic model, the optimal irrigation flow is obtained using numerical methods, with Saint-Venant equations coupled internally with Richards' equation.

The aim of this study is the deduction of an analytic representation of the optimal irrigation flow depending on the border length, hydrodynamic properties, and soil moisture constants, with high values of the coefficient of uniformity.

2. Materials and Methods

2.1. Water Surface Flow

The continuity and amount of movement equations in a border—considering that the effects of the borders were negligible and that water was the shallow or hydraulic hypothesis—were known as equations of Barre de Saint-Venant for border irrigation and written as follows:

$$\frac{\partial h}{\partial t} + \frac{\partial q}{\partial x} + \frac{\partial I}{\partial t} = 0 \tag{1}$$

$$\frac{1}{h}\frac{\partial q}{\partial t} + \frac{2q}{h^2}\frac{\partial q}{\partial x} + \left(g - \frac{q^2}{h^3}\right)\frac{\partial h}{\partial x} + g(J - J_o) + \lambda \frac{q}{h^2}\frac{\partial I}{\partial t} = 0 \tag{2}$$

where $q(x, t) = U(x, t)\, h(x, t)$ is the flow per width unit of border, x is the spatial coordinate in the main direction of movement of water in the border; t is time; U is the mean velocity; h the water depth; J_o is the topographic slope of the border; J is the friction slope; $V_I = \partial I/\partial t$ is the infiltration flow, that is, the volume of water infiltrated in the time unit per width unit and per length unit of the border, I is the infiltrated depth; g is the gravitational acceleration; the dimensionless parameter $\lambda = U_{IX}/U$, with U_{IX} the projection in the movement direction of the output speed of the water body due to infiltration.

The system of equations of Barré de Saint-Venant was not closed since the evolution in time and space of infiltrated depth and friction slope, were unknown. The first was provided by the Richards equation [24] and the second by a law of resistance to the flow that related the friction slope with the hydraulic variables q and h, which were discussed below [2].

2.2. Water Flow in the Soil

If the hypothesis that the irrigation was carried out in flat parallels to the development of the border was accepted, then it is possible to use the two-dimensional form of Richards' equation [24], which results from combining the continuity equation and Darcy's law [33]:

$$C(\psi)\frac{\partial \psi}{\partial t} = \frac{\partial}{\partial x}\left[K(\psi)\frac{\partial \psi}{\partial x}\right] + \frac{\partial}{\partial z}\left[K(\psi)\left(\frac{\partial \psi}{\partial z} - 1\right)\right] \tag{3}$$

where ψ is the potential for water pressure in the soil expressed as the height of an equivalent water column (positive in the saturated zone and negative in the unsaturated zone of soil); $C(\psi) = d\theta/d\psi$ is called the specific capacity of soil moisture; $\theta = \theta(\psi)$ is the water volume per volume unit of soil or water volume content and is a function of ψ, known as the moisture characteristic curve or water retention curve; $K = K(\psi)$ is the hydraulic conductivity, which in a partially saturated soil is a function of the potential of pressure; the gravitational potential is assimilated to the spatial coordinate z positively oriented downwards, x is a spatial coordinate and t is the time.

For the description of the water flow during an irrigation test, the hydrodynamic characterization of soil was necessary. As pointed out by Fuentes et al. [34] in experimental studies, it was more convenient to use the combination of the retention curve proposed by van Genuchten [35], considering the Burdine restriction [36], with the hydraulic conductivity curve proposed by Brooks and Corey [37], due to the fact that they satisfy the integral properties of infiltration and to the ease of identification of their parameters. The retention curve proposed by van Genuchten is written as:

$$\frac{\theta(\psi) - \theta_r}{\theta_s - \theta_r} = \left[1 + \left(\frac{\psi}{\psi_d}\right)^n\right]^{-m} \tag{4}$$

where θ_s is the volumetric water content at effective soil saturation, θ_r is the volumetric content of residual water, ψ_d is a characteristic value of water pressure in the soil, m and n are two parameters of empirical form related here by the Burdine restriction [36]: $m = 1 - 2/n$, with $0 < m < 1$ and $n > 2$.

The hydraulic conductivity proposed by Brooks and Corey [37] is represented as:

$$K(\theta) = K_s\left(\frac{\theta - \theta_r}{\theta_s - \theta_r}\right)^\eta \tag{5}$$

where η is a parameter of positive form whose value can be estimated with $\eta = 2s\,(2/mn + 1)$, being s a function of porosity (ϕ) defined implicitly by $(1 - \phi)^s + \phi^{2s} = 1$ [38].

2.3. Hydraulic Resistance Law

The phase of advance in gravity irrigation is represented by the following initial and boundary conditions in the Barré de Saint-Venant equations:

$$q(x,0) = 0 \quad \text{and} \quad h(x,0) = 0 \tag{6}$$

$$q(0,t) = q_o, \quad q(x_f, t) = 0 \quad \text{and} \quad h(x_f, t) = 0 \tag{7}$$

where $x_f(t)$ is the position of the wave front for the time t and q_o the flow of supply at the entrance of the border.

An analysis of the singularity present in the advance phase in very short times has established that the law of hydraulic resistance, which makes compatible the coupling of Barré de Saint-Venant and Richards equations in this singularity, and it has the following structure [2]:

$$q = kv\left(\frac{h^3 gJ}{v^2}\right)^d \tag{8}$$

where v is the coefficient of kinematic viscosity, k is a dimensionless factor of friction, and d is an exponent such that $1/2 \le d \le 1$ in a way that $d = 1/2$ corresponds to the Chézy turbulent regime and $d = 1$ to the Poiseuille depth regime.

3. Results and Discussion

3.1. Numerical Relationship between the Optimal Flow and Length

The solution of Barre de Saint-Venant and Richards equations to represent surface and subsurface movement, respectively, has been solved numerically based on a Eulerian–Lagrangian scheme that eliminates the traditional instabilities in short times and is available on Saucedo et al. [25].

3.1.1. Optimal Flow in Border Irrigation

With the system of Barré de Saint-Venant and Richards equations, the irrigation design consists of determining the flow of optimal supply and irrigation time to achieve the greatest uniformity along the border, with high levels of application efficiency and irrigation requirement for an irrigation depth and soil hydrodynamic predetermined properties. The optimal flow should be determined for a border length, and its value should be updated in proportion to the new length, which was verified by Rendón et al. [39]. In fact, the optimal flow design follows a linear proportion with the border length that should be applied. The result is obtained using a model formed by the Lewis and Milne [40] equations to describe the water flow on the soil surface and by the Green and Ampt equation [31] to describe the water flow on the soil.

3.1.2. Irrigation Efficiencies

In irrigation, it is essential to distinguish at least three related efficiencies: Application efficiency, irrigation requirement efficiency, and irrigation uniformity efficiency. Application efficiency (η_A) is defined as [1,23]:

$$\eta_A = \frac{V_n}{V_b} = \frac{\ell_n}{\ell_b} \tag{9}$$

where V_n is the water volume required in the root zone of the crop or net volume and V_b is the amount of applied irrigation water. The first is obtained w.ith the expression: $V_n = \ell_n A_r$, where ℓ_n is the net irrigation depth, defined according to the crop irrigation requirements, and A_r is the irrigated area considered. The second is obtained as $V_b = \ell_b A_r$, where ℓ_b is the gross irrigation depth.

Irrigation requirement efficiency (η_R) is defined as [23]:

$$\eta_R = \frac{V_d}{V_n} \tag{10}$$

where V_d is the available volume by the crop. This efficiency indicates how the water needs for the crop are met.

The ideal situation regarding uniformity occurs when all plants receive the same amount of water, a situation which is equivalent to applying the same water depth to the entire length of the border. To evaluate the uniformity in distribution of the infiltrated depth, the Christiansen uniformity coefficient is used. This coefficient (CU_D) results from partitioning the length in N sections of size Δx_i, not necessarily equal, namely [23]:

$$CU_D(t) = 1 - \frac{1}{\bar{I}(t)L}\sum_{i=1}^{N}\left|I_i(t) - \bar{I}(t)\right|\Delta x_i, \quad \bar{I}(t) = \frac{1}{L}\sum_{i=1}^{N}I_i(t)\Delta x_i \tag{11}$$

where I_i is the infiltrated depth at any section i of the border strip or furrow, $\bar{I}$ is the average infiltrated depth, and N is the number of sections considered along the furrow or border strip.

Christiansen classic uniformity coefficient (CUC) results when sections are taken of the same size, $L = N\Delta x$.

3.1.3. Optimal Flow-Length Relationship

The uniformity efficiency measured by the Christiansen uniformity coefficient can be obtained for different combinations of length and supply flow at the head of the border. Saucedo et al. [27] showed an example of four lengths of the border for the soil under study, where it was observed that the uniformity efficiency varied considerably with the irrigation flow.

For each border length, it was possible to determine the value of the supply flow that produced a maximum in the uniformity coefficient with high values of application efficiencies and irrigation requirements. When the supply flow was modified, application efficiencies and irrigation requirements did not vary significantly, not in the case of the uniformity efficiency, which varied substantially with the irrigation flow, i.e., application and irrigation requirement efficiencies can be considered that are not decision variables in defining the optimal flow and, therefore, the uniformity efficiency is that which allows defining the optimal irrigation flow.

The numerical simulation of the irrigation with the system of Barre de Saint-Venant and Richards equations indicates that the relationship between optimal flow (q_o) and length (L) is approximately linear, for a soil type, topographic slope, friction coefficient, and irrigation depth, that is:

$$q_o = q_u L \tag{12}$$

where q_u has units of unitary flow per length unit. In addition, as q_u is a constant, it follows that for the application of a specific irrigation depth, there is an irrigation time, unique and independent of the length, which allows obtaining a maximum value of the uniformity coefficient.

The relation (12) is illustrated by Saucedo et al. [27] in the loam soil of the experimental field of the Colegio de Postgraduados, Montecillo, State of Mexico, for water depths of 8, 10, and 12 cm, by making $q_u = \alpha_u K_s$, where α_u is a dimensionless parameter (Figure 1). It is observed that there is monotony in the sense that the slope of the relationship between the border length and optimal flow decreases as the irrigation depth increases.

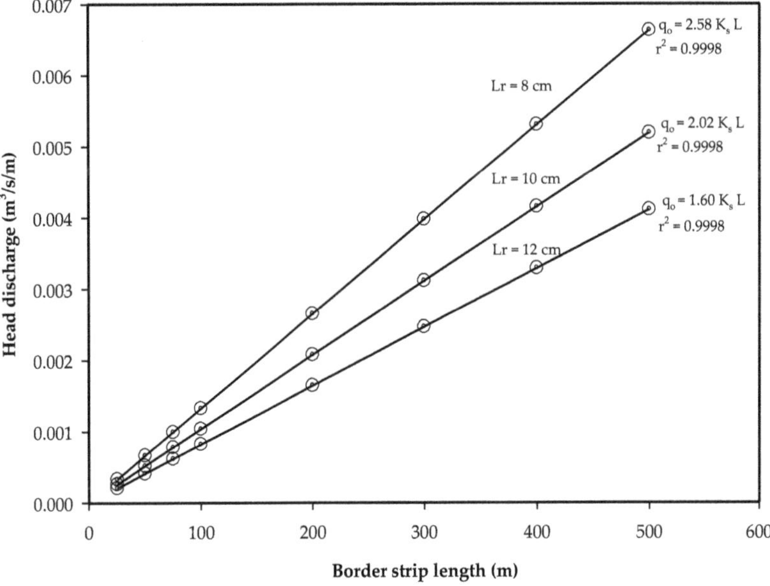

Figure 1. Relationship between the border strip length and the optimum input discharge in the loam soil of Montecillo for three irrigation depths: 8, 10, and 12 cm. K_s in m/s.

3.1.4. Irrigation Design Table

To generate the design table, it was necessary to obtain the relationships between optimal flow and border length for various soil types. The moisture content of residual water (θ_r) was assumed to be zero, according to Fuentes et al. [34]. The moisture content at saturation (θ_s) was assimilated to the total soil porosity (ϕ) determined, as the hydraulic conductivity at saturation (K_s), starting from the soil texture according to the relationships provided by Rawls and Brakensiek [41].

To estimate the value of the shape m parameter of the soil retention curve, one granulometric curve was reconstructed for each soil based on the percentages of sand, silt, and clay present in the triangle of textures [34]; we have followed the procedure suggested by Fuentes [38] to determine the values of m and η. The pressure scale (ψ_d) was determined from the suction of the wetting front of Green and Ampt equation [31] according to the texture and porosity of soils [41], as this suction was identified with the Bouwer scale [42] defined by:

$$\lambda_c = \frac{1}{K_s - K_0} \int_{\psi_0}^{0} K(\psi)d\psi = \frac{1}{K_s - K_0} \int_{\theta_0}^{\theta_s} D(\theta)d\theta \tag{13}$$

where $K_0 = K(\psi_0)$ is the hydraulic conductivity corresponding to the initial pressure $\psi_0 = \psi(\theta_0)$ related to the initial moisture content θ_0, $D(\theta) = K(\theta)\, d\psi/d\theta$ is the hydraulic diffusivity.

The parameter ψ_d is deduced by introducing the hydrodynamic characteristics defined by Equations (4) and (5) in Equation (13), considering the initial moisture content equal to the residual moisture content ($\psi_0 \to \infty$), as follows:

$$\lambda_c = |\psi_d| \frac{1}{n} B\left(\eta m - \frac{1}{n}, \frac{1}{n}\right) \tag{14}$$

where $B(p,q) = \Gamma(p)\Gamma(q)/\Gamma(p+q)$ is the complete beta function, with $p > 0$ and $q > 0$, and $\Gamma(x)$ the Euler complete gamma function.

The parameter values of the hydrodynamic characteristics are shown in Table 1 for different types of soil [27], where the residual moisture content is $\theta_r = 0\ \text{cm}^3/\text{cm}^3$.

Table 1. Hydrodynamic characteristics of soils for irrigation design by borders.

Soil Texture	θ_s $(\text{cm}^3/\text{cm}^3)$	λ_c (cm)	K_s (cm/h)	η	m	$|\psi_d|$ (cm)
Clay	0.525	140.26	0.010	61.10	0.0229	132.50
Silty clay	0.500	100.16	0.015	31.55	0.0440	94.70
Silty-clay-loam	0.500	60.12	0.070	15.34	0.0905	57.80
Clay-loam	0.475	36.00	0.150	19.30	0.0714	34.15
Sandy clay	0.425	25.72	0.200	41.50	0.0327	23.70
Loam	0.500	30.52	0.500	5.61	0.2477	30.70
Silt	0.475	20.04	0.700	13.93	0.0989	19.20
Silty loam	0.525	30.07	0.600	12.01	0.1165	29.35
Sandy-clay-loam	0.425	35.61	1.500	18.44	0.0736	33.35
Sandy loam	0.450	10.00	5.000	13.62	0.1004	9.52

The initial moisture content is considered as that which corresponds when the available moisture of each soil type has been consumed in a certain fraction. The available soil moisture is defined as the difference between the moisture contents at field capacity (θ_{CC}) and permanent wilting point (θ_{PMP}), whose values for each type of soil are estimated according to the soil texture triangle [41]. The initial moisture content is calculated as:

$$\theta_0 = \theta_{PMP} + F_{ap}(\theta_{CC} - \theta_{PMP}) \tag{15}$$

where F_{ap} is the permissible depletion factor of the crop. The average value of 0.5 has been assumed. The values of initial moisture content are reported in Table 2.

Table 2. Moisture constants.

Soil Texture	θ_{PMP} (cm^3/cm^3)	θ_{CC} (cm^3/cm^3)	θ_0 (cm^3/cm^3)
Clay	0.350	0.450	0.400
Silty clay	0.275	0.425	0.350
Silty clay-loam	0.200	0.375	0.287
Clay-loam	0.190	0.340	0.265
Sandy clay	0.225	0.325	0.275
Loam	0.100	0.275	0.187
Silt	0.130	0.250	0.190
Silt loam	0.125	0.275	0.200
Sandy-clay-loam	0.150	0.250	0.200
Sandy loam	0.100	0.190	0.145

As for the parameters of the law of resistance defined by Equation (8), the values were taken from a loam soil border in the Montecillo experimental field. Considering the Reynolds number, the regime is depth, d = 1, the value k = 1/54 is obtained thus that the advance curve provided by the numerical solution describes the advance curve observed in an irrigation test; the coefficient of kinematic viscosity is taken as $v = 10^{-6}$ m^2/s. The longitudinal topographical slope of the border is of $J_0 = 0.002$ m/m, value that is used to simulate irrigation in other borders with different soil types. With the hydrodynamic characterization of soils and θ_0, the constant involved in the relationship between the border length and the optimal flow for a given irrigation depth is calculated.

The value of the constant is expressed in terms of flow per unit area, i.e., per unit width, and per unit length of border, results are shown in Table 3. The same table shows the irrigation time (τ_b) obtained at the moment the flow supply is cut off when the volume of irrigation per width unit is already stored both on the surface as well as inside the soil. In Rendón et al. [39], a table of similar design to Table 3 is shown containing some inconsistencies of monotony between the relationship that the variables optimal flow, irrigation time, and applied depth since the used model has difficulties in reproducing gravity irrigation in relatively long times. The coupling of Saint-Venant and Richards equations allows obtaining results that keep the monotony in the design variables, as shown in Table 3.

Table 3. Table of the border irrigation design: flow in l/s/m^2 for the optimal application of the irrigation depth.

Soil Texture	ℓ_n = 8 cm		ℓ_n = 10 cm		ℓ_n = 12 cm	
	q_u (l/s/m^2)	τ_b (h)	q_u (l/s/m^2)	τ_b (h)	q_u (l/s/m^2)	τ_b (h)
Clay	0.00012	224.1	0.00010	338.2	0.00009	445.0
Silty clay	0.00014	201.6	0.00012	270.5	0.00011	362.5
Silty-clay-loam	0.00060	44.1	0.00050	66.6	0.00046	82.9
Clay-loam	0.00088	31.4	0.00078	44.0	0.00072	57.8
Sandy clay	0.00090	28.7	0.00080	42.4	0.00077	52.0
Loam	0.00399	6.9	0.00333	10.0	0.00296	13.7
Silt	0.00411	6.4	0.00354	9.6	0.00326	12.5
Silt-loam	0.00446	6.2	0.00388	8.8	0.00349	11.6
Sandy-Clay-loam	0.00490	5.8	0.00476	7.4	0.00464	9.0
Sandy-loam	0.02476	1.2	0.02223	1.6	0.02073	2.0

3.2. Analytical Representation of the Relationship between the Optimal Flow and Length

Equation (12) structure is deduced considering that the net water volume per width unit of the border is equal to the product of the border length (L) by the net irrigation depth (ℓ_n) and is also equal to the product of supply unitary flow (q_o) by the time necessary to infiltrate the net depth (τ_n): $q_o\tau_n = L\ell_n$.

The relationship is also attested by involving the irrigation time to obtain the volume per width unit provided by the gross depth: $q_o \tau_b = L\ell_b$. Comparing both results to Equation (12), we have:

$$q_u = \frac{\ell_n}{\tau_n} = \frac{\ell_b}{\tau_b} \tag{16}$$

It should be noted that this relationship involves, considering Equation (9), the following expression for application efficiency:

$$\eta_A = \frac{\ell_n}{\ell_b} = \frac{\tau_n}{\tau_b} \tag{17}$$

From the continuity equation, it can be shown that the unitary flow of minimal irrigation thus that the water wave arrives at the end of the channel is given by $q_m = K_s L$; then it follows that the optimal flow must meet the inequality $q_o \geq q_m$. If $q_o = \alpha_u q_m$ is written, α_u is a dimensionless parameter that must satisfy $\alpha_u \geq 1$. Equations (12) and (16) should be written as follows:

$$q_o = \alpha_u K_s L, \ \alpha_u = \frac{q_u}{K_s} = \frac{\ell_n}{K_s \tau_n} \tag{18}$$

in which the dependence of α_u should be investigated regarding the irrigation depth and soil properties.

The extreme behavior of α_u is deduced from the extreme behavior of the infiltrated depth. In very short times $I = S \sqrt{t}$ [43], where S is sorptivity, and, therefore, $\tau_n = \ell_n^2/S^2$, that is $\alpha_u = S^2/(K_s \ell_n)$. In long times $I \sim I_o + K_s t$, where I_o is the ordinate at the origin depending on S and K_s, and on the Green and Ampt model on the time logarithm, and, therefore, $\alpha_u \sim \ell_n/(\ell_n - I_o)$. From the above, it follows that the limits:

$$\lim_{\ell_n \to 0} \alpha_u = \infty, \ \lim_{\ell_n \to \infty} \alpha_u = 1 \tag{19}$$

must be satisfied by the general function $\alpha_u (\ell_n)$.

Irrigation time (τ_b) shown in Table 3 corresponds to the gross irrigation depth and is greater than the infiltration time corresponding to the net depth calculated from Equation (16): $\tau_n = \ell_n/q_u$. Figure 2 shows the relationship between the two times; one has $\tau_n \approx 0.83 \, \tau_b$ with $r^2 = 0.9995$, which indicates that, according to Equation (17), the average application efficiency with the optimal flow is $\eta_A \approx 83\%$ for analyzed soils.

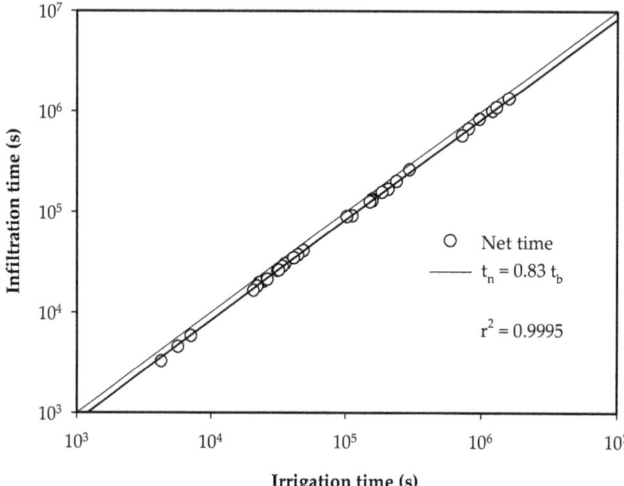

Figure 2. Relationship between the irrigation time of the gross depth (ℓ_b) and the infiltration time of the net depth (ℓ_n).

The design table can be represented algebraically if the infiltration function is provided analytically. This is obtained from the Parlange et al. model [44] deduced from the Richards equation, assuming that the hydraulic diffusivity tends to behave like a Dirac density and form a relationship between the hydraulic diffusivity and hydraulic conductivity. The model with the effect of water depth on the soil surface is as follows [30]:

$$I(t) - K_0 t = \frac{K_s h \Delta\theta}{q_s(t) - K_s} + \frac{S^2}{2\beta(K_s - K_0)} \ln\left[1 + \beta \frac{K_s - K_0}{q_s(t) - K_s}\right], \quad q_s = \frac{dI}{dt} \tag{20}$$

where $\Delta\theta = \theta_s - \theta_0$ is the storage capacity of the soil, h is the water depth on the soil surface, S is the sorptivity, and β is a shape parameter thus that $0 < \beta < 1$, the lower limit corresponds to the Green and Ampt model while the higher limit corresponds to the Talsma and Parlange model [45]. Time in Equation (20) should be interpreted as the contact time that water has at a given point of the border.

Sorptivity S can be calculated with the expression proposed by Parlange [46]:

$$S^2 = \int_{\psi_0}^{0} [\theta_s + \theta(\psi) - 2\theta_0] K(\psi) d\psi \tag{21}$$

and the parameter of β shape can be calculated with the expression proposed by Fuentes [47]:

$$1 - \frac{1}{2}\beta = \frac{\int_{\theta_0}^{\theta_s} \left[\frac{K(\theta)-K_0}{K_s-K_0}\right]\left(\frac{\theta_s-\theta_0}{\theta-\theta_0}\right) D(\theta) d\theta}{\int_{\theta_0}^{\theta_s} D(\theta) d\theta} \tag{22}$$

Variation in time of water depth on the soil is provided by the system of Saint-Venant and Richards equations, but their analytical representation is not known, the reason whereby it is assumed that it is represented by an average value; the mean depth of water can be estimated as a fraction of the normal depth: $\overline{h} = 4/5 \, h_n$. With the dimensionless variables:

$$t^* = \frac{2(K_s - K_0)^2 t}{S^2 + 2K_s \overline{h}\Delta\theta}, \quad I^*(t^*) = \frac{2(K_s - K_0)[I(t) - K_0 t]}{S^2 + 2K_s \overline{h}\Delta\theta} \tag{23}$$

$$q_s^*(t^*) = \frac{dI^*}{dt^*} = \frac{q_s(t) - K_0}{K_s - K_0}, \quad \gamma = \frac{2K_s \overline{h}\Delta\theta}{S^2 + 2K_s \overline{h}\Delta\theta} \tag{24}$$

Equation (20) is written as:

$$I^* = \frac{\gamma}{q_s^* - 1} + \frac{1-\gamma}{\beta} \ln\left[1 + \frac{\beta}{q_s^* - 1}\right] \tag{25}$$

The relationship $dI^*/dq_s^* = q_s^* dt^*/dq_s^*$, considering constant the water depth on the surface, along with the initial condition $t^* = 0$, $I^* = 0$, $q_s^* \to \infty$, leads to find the time as a function of the infiltration flow:

$$t^* = \frac{\gamma}{q_s^* - 1} + \frac{1-\gamma}{\beta(1-\beta)} \ln\left[1 + \frac{\beta}{q_s^* - 1}\right] - \frac{1-\beta\gamma}{1-\beta} \ln\left[1 + \frac{1}{q_s^* - 1}\right] \tag{26}$$

Thus, the function defining the infiltrated depth in the function of time is of a parametric nature: $I^* = I^*(q_s^*)$ and $t^* = t^*(q_s^*)$, with the flow q_s^* as a parameter.

Due to the high nonlinearity of the function $K(\theta)$ it can be assumed that $K_0 \ll K_s$. The dimensionless version of α_u defined in Equation (18) is as follows:

$$\alpha_u = \frac{\ell_n^*\left(q_{sn}^*\right)}{\tau_n^*\left(q_{sn}^*\right)} \tag{27}$$

where the dimensionless net irrigation time and net irrigation depth are defined by Equation (23) and the corresponding dimensionless infiltration flow by Equation (24). For a given border and initial medium depth, the irrigation depth in a dimensionless form (ℓ_n^*) was calculated, and the dimensionless flow (q_{sn}^*) was calculated with Equation (25) iteratively. Finally, the dimensionless net time (τ_n^*) was calculated with Equation (26), and α_u was calculated with Equation (27).

It should be noted that the process of calculating the optimal flow, for a given irrigation length, was also iterative since the medium depth depended on the normal depth and this, in turn, of the optimal flow, through Equation (8): $h_n = [v^2 \, (q_o/kv)^{1/d}/gJ_o]^{1/3}$.

When the water depth was small ($\overline{h} \ll S^2/2K_s\Delta\theta$) from Equation (25) it was deduced an explicit function of time with respect to the infiltrated water and corresponds to the Parlange et al. equation [44]:

$$t^* = I^* - (1-\beta)^{-1} \ln\left\{\beta^{-1}[1-(1-\beta)\exp(-\beta I^*)]\right\} \tag{28}$$

In Table 4, the values of sorptivity and the shape parameter for different soils are reported. In Figure 3, the optimal infiltration time obtained with Saint-Venant and Richards equations is compared with that obtained with Equation (28) of Parlange et al. [44]: $r^2 = 0.9866$.

Table 4. Parameters of the Parlange et al. infiltration equation [44]: sorptivity (S) and the shape parameter (β), calculated with Equations (21) and (29).

Soil Texture	S (cm/$\sqrt{h}$)	β	Soil Texture	S (cm/$\sqrt{h}$)	β
Clay	0.583	0.820	Loam	2.958	0.584
Silty clay	0.655	0.800	Silt	2.761	0.744
Silty-clay-loam	1.296	0.750	Silty loam	3.338	0.721
Clay loam	1.469	0.773	Sandy-clay-loam	4.796	0.775
Sandy clay	1.223	0.820	Sandy loam	5.404	0.744

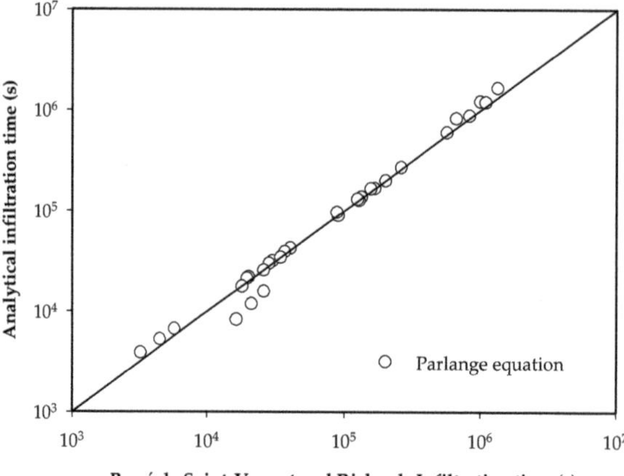

Figure 3. Comparison between the optimal irrigation time numerically obtained with the Barré de Saint-Venant/Richards equations and those calculated ones with the Parlange et al. equation (Equation (28)).

The shape parameter β varies with respect to the initial moisture content; however, it does not vary significantly when this moisture content approaches the residual moisture content. In this case the introduction of the hydrodynamic characteristics defined by Equations (4) and (5) in Equation (22) provides:

$$\beta = 2\left\{1 - \frac{B[(2\eta - 1)m - 1/n, 1/n]}{B(\eta m - 1/n, 1/n)}\right\} \tag{29}$$

where $B(p,q)$ is the complete beta function.

Introduction of Equation (28) into Equation (27) gives the approximate formula for calculating the optimal unitary flow in function of the border length, the net depth, and the characteristic parameters of infiltration representing the capillary forces (sorptivity) and gravitational forces (hydraulic conductivity at saturation) and a shape parameter of the hydrodynamic characteristics, namely:

$$\alpha_u = \frac{\ell_n^*}{\ell_n^* - (1 - \beta)^{-1} \ln\left\{\beta^{-1}[1 - (1 - \beta)\exp(-\beta\ell_n^*)]\right\}}, \quad \ell_n^* = \frac{2K_s\ell_n}{S^2} \tag{30}$$

where ℓ_n^* is the net irrigation depth in dimensionless writing.

As can be seen, Equation (30) contains a shape parameter in the function of the soil type; however, it does not have a great variation in the range of soils reported in Table 4, the mean value $\beta \approx 3/4$ can be taken. With this value in Figure 4, the graph of Equation (30) and its comparison with the numerical results are presented. A good agreement is clearly demonstrated.

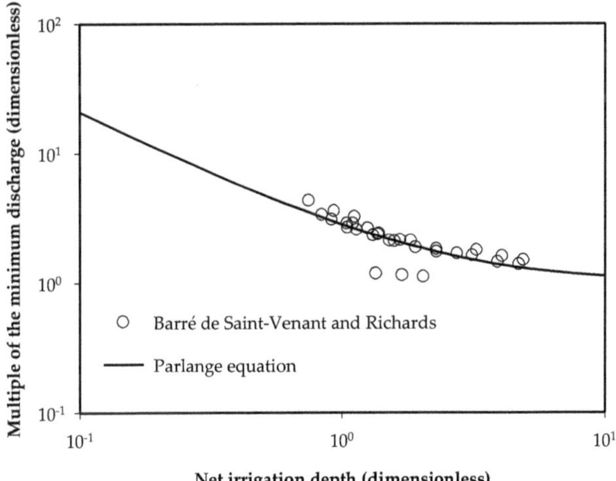

Figure 4. The multiple factors of the minimum discharge as a function of the irrigation depth numerically obtained with the Saint-Venant and Richards equations, and those calculated with the Parlange et al. model [44], Equation (30).

3.3. Application of the Analytical Formula in Furrow Irrigation Systems

This analytical formula has been applied in field experiments realized by Chávez and Fuentes [1] with good results. Irrigation tests (250) were performed in 1010 ha with the next crops: *Zea mayz, Sorghum vulgare, Medicago sativa, Phaseolus vulgaris, Pachyrhizus erosus, Hordeum vulgare, Triticum aestivum,* and *Allium cepa.*

The characteristics and properties that were measured in the plots were: Length, slope, texture, bulk density, initial, and saturation moisture content. The results that were obtained from the evaluation of the irrigation tests were: Discharge at the entrance of the plot, number of furrows by irrigation set, saturated hydraulic conductivity, the efficiency of the application, and the slope of the direction of irrigation. The irrigation tests were performed in plots.

With the advanced and recession data and the characteristics of the soils from the locations where the irrigation tests were carried out, the calibration of the test was performed using the kinematic wave model. With the parameters found (K_s and h_f) for the evaluation of the irrigation tests and the net irrigation depth that is intended to be applied on the plot (water depth depending on each of the crops established in the plots), Equations (18) and (30) were used to make the design. The obtained result is the optimal flow that must go into each furrow; for this value, the discharge for the entrance of the plot between that obtained with Equation (18) was divided, and then approached the nearest whole value. As a result, the equation gives us the number of furrows that the user has to open by set and the time that must pass before cutting off the water.

In this study, the evaluation of irrigation tests, the data of the plot, and the net irrigation depth to be applied demonstrated that the optimal flow expense that can be put in each furrow during an irrigation event can be calculated under the hypothesis that with this expense, the historical water depths applied in the evaluated plots can be reduced. The average water depths decreased by 19 cm, irrigation time decreased 12 h ha^{-1} on average, and the average volume saved was 2150 m^3 ha^{-1}, which represented a total of 49% of the total volume used. In addition, the average efficiency rose from 51% to 86%.

4. Conclusions

A linear relationship has been validated between the length of the border and the optimal irrigation flow, defined as the inflow rate that has to be applied to obtain a maximum value of Christiansen's uniformity coefficient with high values of application efficiency and irrigation requirement efficiency. The linear form of the proportion between both variables was corroborated by [39] using a hydrological model for the flow of surface water and the Green and Ampt infiltration equation.

The proportion between optimal flow and border length has been established by numerical simulation with the hydrodynamic model, formed by the coupled equations of Barré de Saint-Venant and Richards. In the numerical simulation, 10 types of soils of contrasting texture have been used and three water depths applied, which has allowed us to form an irrigation design table, that is, 10 linear relationships for each irrigation depth.

To establish an analytical representation of the optimal flow in function of the border length, the Parlange infiltration theory proposed for integrating the differential equation of the vertical one-dimensional infiltration has been used. In the formula obtained to relate the optimal irrigation flow and length, the irrigation depth intervenes and as soil parameters sorptivity that comes from the capillary forces and the saturated hydraulic conductivity that comes from gravitational forces. The good accordance between the results of numerical simulation and analytical representation allows us to obtain the formula for the design of gravity irrigation.

Author Contributions: Conceptualization, C.F.; methodology, C.F. and C.C.; software, C.F.; validation, C.C. and C.F.; formal analysis, C.C.; investigation, C.F.; resources, C.F. and C.C.; data curation, C.F. and C.C.; writing—original draft preparation, C.F.; writing—review and editing, C.C.; visualization, C.F. and C.C.; supervision, C.F.; project administration, C.F. and C.C.; funding acquisition, C.F. All authors have read and agreed to the published version of the manuscript.

Funding: This research received no external funding.

Conflicts of Interest: The authors declare no conflict of interest.

References

1. Chávez, C.; Fuentes, C. Design and evaluation of surface irrigation systems applying an analytical formula in the irrigation district 085, La Begoña, Mexico. *Agric. Water Manag.* **2019**, *221*, 279–285. [CrossRef]
2. Fuentes, C.; De León, B.; Saucedo, H.; Parlange, J.-Y.; Antonino, A.C.D. Saint-Venant and Richards equations system in surface irrigation: (1) Hydraulic resistance power law. *Ing. Hidraul. Mexico* **2004**, *19*, 65–75.

3. Khanna, M.; Malano, H.M. Modelling of basin irrigation systems: A review. *Agric. Water Manag.* **2006**, *83*, 87–99. [CrossRef]

4. Soroush, F.; Fenton, J.D.; Mostafazadeh-Fard, B.; Mousavui, S.F.; Abbasi, F. Simulation of furrow irrigation using the Slow-change/slow-flow equation. *Agric. Water Manag.* **2013**, *116*, 160–174. [CrossRef]

5. Ebrahimian, H.; Liaghat, A. Field Evaluation of Various Mathematical Models for Furrow and Border Irrigation Systems. *Soil Water Res.* **2011**, *2*, 91–101. [CrossRef]

6. Furman, A.; Warrick, A.W.; Zerihun, D.; Sanchez, D.A. Modified Kostiakov infiltration function: Accounting for initial and boundary conditions. *J. Irrig. Drain. Eng.* **2006**, *132*, 587–597. [CrossRef]

7. Parhi, P.K.; Mishra, S.K.; Singh, R. A modification to Kostiakov and modified Kostiakov infiltra-tion models. *Water Resour. Manag.* **2007**, *21*, 1973–1989. [CrossRef]

8. Zhang, S.; Xu, D.; Li, Y. A one-dimensional complete hydrodynamic model of border irrigation based on a hybrid numerical method. *Irrig. Sci.* **2011**, *29*, 93–102. [CrossRef]

9. Strelkoff, T.; Katopodes, N. Border-irrigation hydraulics with zero inertia. *J. Irrig. Drain. Div. ASCE* **1977**, *102*, 325–342.

10. Bautista, E.; Wallender, W.W. Hydrodynamic furrow irrigation model with especified space steps. *J. Irrig. Drain. Eng. Div. ASCE* **1992**, *118*, 450–465. [CrossRef]

11. Sohrabi, B.; Behnia, A. Evaluation of Kostiakov´s infiltration equation in furrow irrigation design according to FAO Method. *J. Agron.* **2007**, *6*, 468–471.

12. Bautista, E.; Clemmens, A.J.; Strelkoff, T.S.; Schlegel, J. Modern analysis of surface irrigation systems with WinSRFR. *Agric. Water Manag.* **2009**, *96*, 1146–1154. [CrossRef]

13. Strelkoff, T.; Clemmens, A.J. Hydraulics of surface systems. In *Design and Operation of Farm Irrigation Systems*, 2nd ed.; Hoffman, G.J., Evans, R.G., Jensen, M.E., Martin, D.L., Elliot, R.L., Eds.; ASABE: St. Joseph, MO, USA, 2007; pp. 436–498.

14. Gillies, M.H.; Smith, R.J. SISCO: Surface irrigation simulation, calibration and optimisation. *Irrig. Sci.* **2015**, *33*, 339–355. [CrossRef]

15. Valipour, M. Comparison of surface irrigation simulation models: Full hydrodynamic, zero inertia, kinematic wave. *J. Agric. Sci.* **2012**, *4*, 68–74. [CrossRef]

16. Mahdizadeh, K.M.; Gholami, S.M.A.; Valipour, M. Simulation of open- and closed-end border irrigation systems using SIRMOD. *Arch. Agron. Soil Sci.* **2015**, *61*, 929–941. [CrossRef]

17. Salahou, M.K.; Jiao, X.; Lü, H. Border irrigation performance with distance-based cut-off. *Agric. Water Manag.* **2018**, *201*, 27–37. [CrossRef]

18. Akbari, M.; Gheysari, M.; Mostafazadeh-Fard, B.; Shayannejad, M. Surface, irrigation simulation-optimization model based on meta-heuristic algorithms. *Agric. Water Manag.* **2018**, *201*, 46–57. [CrossRef]

19. Lalehzari, R.; Nasab, S.B. Improved volume balance using upstream flow depth for advance time estimation. *Agric. Water Manag.* **2017**, *186*, 120–126. [CrossRef]

20. Liu, K.H.; Jiao, X.Y.; Li, J.; An, Y.H.; Guo, W.H.; Salahou, M.K.; Sang, H. Performance of a zero-inertia model for irrigation with rapidly varied inflow discharges. *Int. J. Agric. Biol. Eng.* **2020**, *13*, 175–181.

21. Moravejalahkami, B.; Mostafazadeh-Fard, B.; Heidarpour, M.; Abbasi, F. Research Paper: SW-Soil and Water Furrow infiltration and roughness prediction for different furrow inflow hydrographs using a zero-inertia model with a multilevel calibration approach. *Biol. Eng.* **2009**, *103*, 374–381.

22. González, C.J.M.; Muñoz, H.B.; Acosta, H.R.; Mailhol, J.C. Kinematic wave model adapted to irrigation with closed-end furrows. *Agrociencia* **2006**, *40*, 731–740.

23. Rendón, P.L.; Saucedo, H.; Fuentes, C. *Diseño del Riego por Gravedad*, 1st ed.; Fuentes, C., Rendón, L., Eds.; Universidad Autónoma de Querétaro: Santiago de Querétaro, Mexico, 2012; pp. 321–358.

24. Richards, L.A. Capillary conduction of liquids through porous mediums. *Physics* **1931**, *1*, 318–333. [CrossRef]

25. Saucedo, H.; Fuentes, C.; Zavala, M. The Saint-Venant and Richards equation system in surface irrigation: (2) Numerical coupling for the advance phase in border irrigation. *Ing. Hidraul. Mexico* **2005**, *20*, 109–119.

26. Saucedo, H.; Zavala, M.; Fuentes, C. Complete hydrodynamic model for border irrigation. *Water Technol. Sci.* **2011**, *2*, 23–38.

27. Saucedo, H.; Zavala, M.; Fuentes, C.; Castanedo, V. Optimal flow model for plot irrigation. *Water Technol. Sci.* **2013**, *4*, 135–148.

28. Castanedo, V.; Saucedo, H.; Fuentes, C. Comparison between a hydrodynamic full model and a hydrologic model in border irrigation. *Agrociencia* **2013**, *47*, 209–223.

29. Schmitz, G.; Haverkamp, R.; Palacios, O. A coupled surface-subsurface modelor shallow water flow over initially dry-soil. In Proceedings of the 21st Congress (IAHR), Melbourne, Australia, 13–18 August 1985; Volume 1, pp. 23–30.

30. Parlange, J.-Y.; Haverkamp, R.; Touma, J. Infiltration under ponded conditions. Part I. Optimal analytical solutions and comparisions with experimental observations. *Soil Sci.* **1985**, *139*, 305–311. [CrossRef]

31. Green, W.H.; Ampt, G.A. Studies in soil physics, I: The flow of air and water through soils. *J. Agric. Sci.* **1911**, *4*, 1–24.

32. Vázquez, F. Eficiencia de aplicación en el riego con surcos cerrados al existir dos pendientes. *Ing. Investig. Tecnol.* **2000**, *3*, 123–130.

33. Darcy, H. Dètermination des lois d'ècoulement de l'eau à travers le sable. In *Les Fontaines Publiques de la Ville de Dijon*; Victor Dalmont: Paris, France, 1856; pp. 590–594.

34. Fuentes, C.; Haverkamp, R.; Parlange, J.-Y. Parameter constraints on closed–form soil-water relationships. *J. Hydrol.* **1992**, *134*, 117–142. [CrossRef]

35. Van Genuchten, M.T. A closed-form equation for predicting the hydraulic conductivity of unsaturated soils. *Soil Sci. Soc. Am. J.* **1980**, *44*, 892–898. [CrossRef]

36. Burdine, N.T. Relative permeability calculation from size distributions data. *Trans. AIME* **1953**, *198*, 171–199. [CrossRef]

37. Brooks, R.H.; Corey, A.T. Hydraulic properties of porous media. *Hydrol. Pap. (Colo. State Univ.)* **1964**, *3*, 27.

38. Fuentes, C. Approche Fractale des Transferts Hydriques Dans les sols non Saturès. Ph.D. Thesis, Universidad Joseph Fourier de Grenoble, Francia, France, 1992; p. 267.

39. Rendón, P.L.; Saucedo, H.; Fuentes, C. Gravity Irrigation Design. In *Gravity Irrigation*, 1st ed.; Fuentes, C., Rendón, L., Eds.; National Association of Irrigation Specialist: Mexico, Mexico, 2017; pp. 345–386.

40. Lewis, M.R.; Milne, W.E. Analysis of border irrigation. *Agric. Eng.* **1938**, *19*, 267–272.

41. Rawls, W.J.; Brakensiek, D.L. Estimating soil water retention from soil properties. *Am. Soc. Civ. Eng.* **1981**, *108*, 167–171.

42. Bouwer, H. Rapid field measurement of air entry value and hydraulic conductivity of soil as significant parameters in flow system analysis. *Water Resour. Res.* **1964**, *36*, 411–424. [CrossRef]

43. Philip, J.R. The theory of infiltration: 1. The infiltration equation and its solutions. *Soil Sci.* **1957**, *83*, 345–357. [CrossRef]

44. Parlange, J.-Y.; Braddock, R.D.; Lisle, I.; Smith, R.E. Three parameter infiltration equation. *Soil Sci.* **1982**, *111*, 170–174. [CrossRef]

45. Talsma, T.; Parlange, J.-Y. One-dimensional vertical infiltration. *Aust. J. Soil. Res.* **1972**, *10*, 143–150. [CrossRef]

46. Parlange, J.-Y. On solving the flow equation in unsaturated soils by optimization: Horizontal infiltration. *Soil Sci. Soc. Am. Proc.* **1975**, *39*, 415–418. [CrossRef]

47. Fuentes, C. Teoría de la infiltración unidimensional: 2. La infiltración vertical. *Agrociencia* **1989**, *78*, 119–153.

© 2020 by the authors. Licensee MDPI, Basel, Switzerland. This article is an open access article distributed under the terms and conditions of the Creative Commons Attribution (CC BY) license (http://creativecommons.org/licenses/by/4.0/).

 *water*

Article

The Effect of Irrigation Treatment on the Growth of Lavender Species in an Extensive Green Roof System

Angeliki T. Paraskevopoulou [1,*], Panagiotis Tsarouchas [1], Paraskevi A. Londra [2] and Athanasios P. Kamoutsis [3]

[1] Laboratory of Floriculture and Landscape Architecture, Department of Crop Science, School of Plant Sciences, Agricultural University of Athens, Iera Odos 75, 118 55 Athens, Greece; pan.tsarouchas@gmail.com

[2] Laboratory of Agricultural Hydraulics, Department of Natural Resources Management and Agricultural Engineering, School of Environment and Agricultural Engineering, Agricultural University of Athens, Iera Odos 75, 118 55 Athens, Greece; v.londra@aua.gr

[3] Laboratory of General and Agricultural Meteorology, Department of Crop Science, School of Plant Sciences, Agricultural University of Athens, Iera Odos 75, 118 55 Athens, Greece; akamoutsis@aua.gr

* Correspondence: aparas@aua.gr; Tel.: +30-210-529-4551

Received: 30 January 2020; Accepted: 17 March 2020; Published: 19 March 2020

Abstract: In green roofs, the use of plant species that withstand dry arid environmental conditions and have reduced water requirements is recommended. The current study presents the effect of irrigation amount on the growth of four different species of lavender; *Lavandula angustifolia*, *Lavandula dentata* var. *candicans*, *Lavandula dentata* var. *dentata*, and *Lavandula stoechas* established on an extensive green roof system and used in urban agriculture. Two irrigation treatments (high and low) determined by the substrate hydraulic properties were applied. Plant growth studied at regular intervals included measurements of plant height, shoot canopy diameter, plant growth index, shoot dry weight and stomatal conductance. The results were consistent and showed that low irrigation reduced plant growth. With the exception of *L. stoechas*, the appearance of plants watered with the low irrigation treatment was satisfactory, and their use under low water amount irrigation is supported. Interspecies differences among lavender species were present in both irrigation treatments. Overall, *L. dentata* var. *candicans* showed the greatest growth, followed in descending order by *L. dentata* var. *dentata* and *L. angustifolia*. In parallel, for stomatal conductance, *L. dentata* var. *candicans* showed the lowest value, similar to *L. dentata* var. *dentata*, and *L. angustifolia* the largest. Differences in plant characteristics and size among the latter three species can be considered in the design of extensive green roof systems. The use of substrate hydraulic properties was shown to be important for irrigation management on extensive green roof systems.

Keywords: drought tolerance; substrate hydraulic properties; substrate available water; ornamental; aromatic; urban agriculture; green infrastructures

---

1. Introduction

There is increasing interest in urban agriculture due to the related economic, social and environmental functions contributing to the sustainability of cities [1]. Though urban agriculture usually highlights food production, it includes the cultivation of other plants such as ornamentals [2], as well as agricultural systems that relate to recreation and leisure [3]. However, within cities, land and soil are limited resources [4]. Generally, cities are characterised by dense buildings, green spaces which are limited in number and size and large impervious paved areas. These characteristics have contributed to creating adverse environmental conditions within the cities such as the heat island effect, restricted air flows, human discomfort and poor health caused by heat stress and poor air quality [5]. Roof greening (the development of planting on buildings, i.e., green roofs) is one means by which urban agriculture may be realized [1,6]; it has the potential to contribute to mitigating the problems caused by urbanisation on an individual scale, and when applied broadly, could improve the environment of a city [5].

Green roofs are generally classified into three categories depending on weight, substrate layer, maintenance, cost, plant community, and irrigation, i.e., extensive, semi-intensive and intensive roofs [7]. Within cities, the load-bearing capacity of many buildings, particularly older ones, is limited; hence, only extensive green roof systems can be applied on these buildings due to their smaller weight load in comparison to other green roof systems. Extensive green roofs are characterised by shallow depths and reduced water availability. Water is an additional limited natural resource within many cities, particularly in semiarid and arid locations such as the Mediterranean, and especially during the summer months [8]. Furthermore, in Mediterranean regions, high temperatures make the development of green roofs more difficult [9]. Under the increasing threat of climate change, water conservation is a priority. Therefore, it is critical in extensive green roofs to use plant species that withstand dry heat and water-deficits [10]. In recent years, research on the growth of shrubs in extensive green roofs is increasing [8,11–16]. Plant growth in extensive green roofs with limited irrigation was found to be satisfactory for *Artemisia absinthium* L., *Helichrysum italicum* Roth., *Helichrysum orientale* L. [11,12], *Origanum majorana* L., and *Santolina chamaecyparissus* L. [12] at a substrate depth of 7.5 cm, and for *Arthrocnemum macrostachyum*, *Halimione portulacoides* [8], *Convolvulus cneorum* L. [13,14], *Origanum dictamnus* L. [14], *Atriplex halimus* [16], and *Pallenis maritima* [15] at a substrate depth of 10 cm.

Generally, the literature on drought-resistant plants for use in agriculture and landscape architecture is extensive [17]. On the other hand, there is a need to study the survival of shrubs on green roofs in hot and dry climates [10]. The amount of water loss in extensive green roofs is a function of three properties of the green roof system, i.e., plant water uptake and transpiration, shading of substrate by vegetation that might reduce the substrate surface evaporation rate and greater water holding capacity of the substrates containing plant roots [18]. A balance among species of water and substrate is needed to address the adverse environmental conditions of green roofs and the effect of temperature extremes [19]. Therefore, plant selection and the improvement of the available amount of water to plants are key research aims [9]. Plant survival on green roofs with shallow substrates and low water availability is not easily understood, and is determined by a combination of drought avoidance physiological processes [10] such as the decline of stomatal conductance, and hydraulic conductivity [20] expressed by species in various ways that include dormancy, drought deciduousness and stomatal regulation [10]. Several authors believe that the first response of plants to severe drought is the closure of their stomata to prevent transpiration water loss [21–23]. Species that are well adapted to drought, such as *Olea europaea* L., decrease water loss through stomatal closure from early in the morning [24]. Stomatal conductance plays an essential role in regulating plant water balance and may reduce plant transpiration [23]; however, it may also concomitantly reduce cell expansion and growth rate, leading to reduced biomass and yield [22,25].

In accordance with De Boodt and Verdonck [26], plant growth decreases when water retention in substrates occurs at negative pressure heads greater than −100 cm, and inadequate substrate aeration conditions for plant growth are created when negative pressure heads are less than −10 cm. Therefore, retaining substrate water content within the available water range defined by negative pressure heads between −10 and −100 cm during irrigation ensures substrate water availability and plant water uptake, thereby reducing the effect of water stress. A comparative study for investigating plant growth among different lavender species (family: Lamiaceae) on a simulated extensive green system and under different irrigation treatments determined by the hydraulic properties of the substrate has not been undertaken before. Lamiaceae is characterised by numerous aromatic species of arid and warm climates. The *Lavandula* genus includes 47 species and many varieties [27]. The qualitative characteristics of the different species such as the habit and morphological characteristics of the flowers and foliage vary [28,29], and are of interest to both landscape and urban agriculture. In this study, the hydraulic properties of an extensive green roof system substrate were used to determine different amounts of irrigation within the available water range defined by negative pressure heads between −10 and −100 cm. The objective of this study was to investigate the effect of irrigation amount on the growth of 4 lavender species, i.e., *Lavandula angustifolia*, *Lavandula dentata* var. *candicans*, *Lavandula dentata* var. *dentata* and *Lavandula stoechas* on an extensive green roof system under two irrigation treatments (high and low) in the aforementioned available water range to support the creation of aesthetically-pleasing green roofs for urban agriculture.

2. Materials and Methods

2.1. Experimental Setup

Four popular lavender species were selected for study, *Lavandula angustifolia*, *Lavandula dentata* var. *dentata*, *Lavandula dentata* var. *candicans* and *Lavandula stoechas*. Uniform, 9 cm size pot lavender plants were supplied by the Kalantzis Plants (Marathonas, Greece) nursery. Plants were individually transplanted on 1 March 2016 in rectangular shaped 60 cm × 40 cm plastic containers (1 plant per container), simulating an extensive green roof system comprised bottom-up from a water retention and protection layer, a drainage layer, a filter layer, and 10 cm deep substrate [8]. The substrate used was S_{15}:Pum_{70}:C_{15} and consisted of soil (S), pumice (Pum) and grape marc compost (C) in a volumetric ratio of 15:70:15. Containers were positioned on metal benches (0.80 m height) on the roof of the main building of the Agricultural University of Athens (lat. 37°58′57″ N, long. 23°42′17″ E, alt. 30 m) to avoid the effect of shading from the perimeter walls of the roof. After transplanting, plants were left to grow and establish for 3 months (1 March—30 May 2016). An automated irrigation system was applied using a drip system with two emitters of 2L h^{-1} per plant spaced at 10 cm on either side of the plant and a total irrigation water application rate of approximately 16.6 mm h^{-1}. Throughout the study period on a monthly basis, 1.2 g L^{-1} H_2O Nutri-Leaf 20–20–20 (Miller Chemical and Fertilizer Corporation, U.S.A.) of fertilizer was applied to all plants. During the experiment, there was no leaching from the applied fertilizer, as the water application rate was gradual and the applied irrigation amount produced no water excess (see *2.3. Experimental Design and Irrigation Treatments*). The duration of the experiment was 4 months and took place mainly over the summer months, from 31 May (day 1) to 30 September 2016 (day 123).

2.2. Physical-Hydraulic Properties of Substrate

The S_{15}:Pum_{70}:C_{15} substrate had a bulk density $\rho_\varphi = 1.035$ g cm^{-3}, pH = 7.8 and EC = 1.33 dS m^{-1} (the latter two measurements were made in 1:1 solution extract). The soil used was sandy loam/loam (53.62% sand, 30.82% silt, 15.56% clay, 0.7% organic matter), the pumice contained particles of diameter size 0.06–8 mm (LAVA, Mining & Quarrying A.D, Athens, Greece) and the grape marc compost (i.e., a waste product of wine production) was composted for 20 months and used as a sustainable alternative to peat. The particle size distribution of the substrates was determined with screen analysis. Weighed substrate samples were placed in the top sieve of a column of sieves arranged from top to bottom in descending order of screen mesh size (>20.00, 16.00, 10.00, 8.00, 4.00, 2.00, 1.00, 0.50, 0.25, 0.106, and <0.053 mm) resting on a sieve shaker for 3 min at 30 shakes per minute.

A tension plate apparatus in a Haines-type assembly [30], with an air-entry value of −180 cm of a water column was employed to define the substrate water retention curve. The substrate sample of 3 cm in height and 10.2 cm in diameter was positioned on the vibrating porous plate of a Buchner filter funnel to achieve satisfactory packing. It was then subjected to gradual wetting from the bottom of the plate until saturation (for 48 h). Measurements of the water content at different pressure heads were taken to obtain the water retention curve. The retention curve was the mean of three substrate samples (n = 3).

The RETC program [31] was used to calculate the fitting hydraulic parameters of the widely used Mualem-van Genuchten model [32,33] on the experimental water retention data. Van Genuchten [33] described the water retention curve as

$$\theta(H) = (\theta_s - \theta_r)\left(\frac{1}{1 + (\alpha|H|)^n}\right)^m + \theta_r,$$

(1)

where θ denotes the soil water content (cm^3 cm^{-3}), subscripts s and r denote the saturated and residual values of water content, α is the curve-fitting parameter inversely proportional to the mean pore diameter (cm^{-1}), and both m, n are dimensionless shape curve-fitting parameters, $m = 1 - 1/n$ and $0 < m < 1$.

Combining Equation (1) with the model developed by Mualem [32], the relationship between hydraulic conductivity and soil water content, $K(\theta)$, can be calculated as

$$K(\theta) = K_s\left(\frac{\theta - \theta_r}{\theta_s - \theta_r}\right)^{0.5}\left\{1 - \left[1 - \left(\frac{\theta - \theta_r}{\theta_s - \theta_r}\right)^{1/m}\right]^m\right\}^2,$$

(2)

The model fitting parameters described above were evaluated by the RETC program using the measured water retention and saturated hydraulic conductivity data. The unknown parameters of the Mualem-van Genuchten model in the parameter optimization process to fit the water retention function were α, n and θ_r.

The saturated hydraulic conductivity, K_s, was determined by the constant-head method [34].

2.3. Experimental Design and Irrigation Treatments

The effect of the amount of irrigation water on the plant growth of the four selected lavender species (*Lavandula angustifolia*, *Lavandula dentata* var. *dentata*, *Lavandula dentata* var. *candicans* and *Lavandula stoechas*) was studied. The plant containers were arranged in a randomised design with 6 replicates per species. The amount of irrigation water was based on the substrate available water defined by the water retention curve of the substrate (i.e., water content released between −10 and −100 cm pressure head). Two irrigation treatments, i.e., high and low amounts of water, were applied through the automated irrigation system using two irrigation 9001 controllers (Galgon, Kfar Blum, Israel). Plants irrigated with a high amount of water were not subjected to water stress and served as the control. The above irrigation treatments were applied for 4 months from day 1 of the experiment (31 May 2016) until day 123 (30 September 2016). Substrate water content was measured daily using a handheld Frequency Domain Reflectometry (FDR) soil moisture sensor (HH2, Delta-T Devise, Cambridge, U.K.; WET Sensor type WET-2, Delta-T Devise, Cambridge, U.K.) set at the 'mineral' setting and calibrated to the used substrate. The sensor was fully inserted into the substrate with the central rod positioned 5 cm away from the plant centre. During the high irrigation treatment, when the FDR sensor showed a water content value of approximately 0.31 cm^3 cm^{-3} (at corresponding pressure head −50 cm), the plants were irrigated with 1.95 L, ensuring water availability within the easily available water (EAW) area. In the case of the low irrigation treatment, plants were irrigated with half the amount of the high irrigation treatment, i.e., 0.975 L, when the FDR sensor showed a water content value of approximately 0.29 cm^3 cm^{-3} (at corresponding pressure head −100 cm).

2.4. Plant Growth Biometrics

On day 1, plant size (height and shoot canopy diameter) was similar among species. Plant height (determined from the pot rim of the substrate surface), shoot canopy diameter (average of the widest and perpendicular to the widest plant diameter), and growth index [(height + widest width + perpendicular width)/3] were measured at monthly intervals. In all plants, at the end of the experiment (30 September/ day 123), stomatal conductance was recorded on the abaxial surface of the third or fourth fully expanded leaf from the stem base on the exterior of the plant using the AP4 Porometer (Delta-T Devices, Cambridge UK). Measurements were taken between 13:00–14:00 h and three readings were recorded per plant and averaged. Next, in all plants, shoots were individually harvested at the end of the experiment (day 123). In all species and both irrigation treatments, plant roots penetrated the substrate and could not be separated from the substrate without partial loss of the fine root system; therefore, it was decided that the plant roots would not be harvested. The harvested shoots were dried in an oven at 70 °C for 48 h, and their dry weights were determined. Finally, in all plants, the percentage increase in plant height, shoot canopy diameter and growth index (GI) was calculated by dividing the difference between the last and first corresponding measurement with the first corresponding measurement and then multiplying by a hundred. Weekly recordings of observations for potential signs of water stress were undertaken throughout the duration of the experiment. In plants, the onset of visual symptoms induced by drought (leaf and stem chlorosis and necrosis) was recorded during the experiment and assessed at the end (on day 123). Visual symptoms induced by irrigation treatments were assessed on a 6-point scale from 0–5, where 0: plant mortality, 1: very severe leaf rolling and chlorosis >75%, 2: severe leaf rolling and chlorosis 50–75% approximately, 3: moderate leaf rolling and chlorosis 25–50% approximately, 4: mild leaf rolling and chlorosis <25%, 5: no leaf injury.

2.5. Meteorological Conditions

Meteorological data were obtained by the nearby meteorological station at Thissio (lat. 38°0.00′ N long. 23°43.48′ E, alt. 110m) of the National Observatory of Athens, located 1.8 km away from the experimental site [35]. In 2016, the average daily temperatures ranged between 19.9 °C on 25 September, and 32.8 °C on 21 June, while the absolute maximum and minimum air temperature values of 39.9 °C and of 16.5 °C were recorded on 21 June and 26 September, respectively (Figure 1). Diurnal temperature range (T_{max}–T_{min}) fluctuated between 4.4–14.1 °C with an average of 8.8 °C. In more detail, during the initial period of the experiment, i.e., 31 May (day 1) until 30 June (day 31), warm thermal conditions dominated, ranging from 17.3 °C to 39.8 °C. In August, air temperature continued to fluctuate from 21.8 °C to 38.5 °C. A decrease in air temperature was observed during the end of the experimental period—from 1 September (day 94) to 30 September (day 123)—with fluctuations from 16.5 °C to 32.9 °C. Overall, July (day 32–62) and August (day 63–93) were the hottest months during the experiment.

Throughout the experiment period, there was very little precipitation concentrated near the start (4 days in June) and end (6 days in September) (Figure 1). More specifically, on 7 and 28 June, 2016 precipitation was approximately 7.6 and 9.6 mm, respectively, and on all the other rainy days, the precipitation was <2 mm. Furthermore, there was no rainfall in either July or August 2016. During the experiment, both July and August showed the least relative humidity (Figure 1). Hence, the hottest and least humid period of the experiment took place during days 33–93 (1 July and 31 August 2016). Furthermore, the air temperature values were higher, while the precipitation and relative humidity values were less than the corresponding climatic (normal) values (reference period 1961–1990). Therefore, during the time of the experiment, more hot and dry conditions than the climatic values prevailed [35]. Furthermore, measurements of meteorological stations are typically representative within a 10 km radius [36]. In the current study, the experimental site was located 1.8 away from the meteorological station; therefore, the obtained meteorological data are representative. Microclimate conditions of the experimental site were not recorded, however, due to the adverse environmental conditions of the extensive green roofs [22], and it is likely that the ambient temperate and relative humidity levels were greater and less than the corresponding data obtained from the meteorological station.

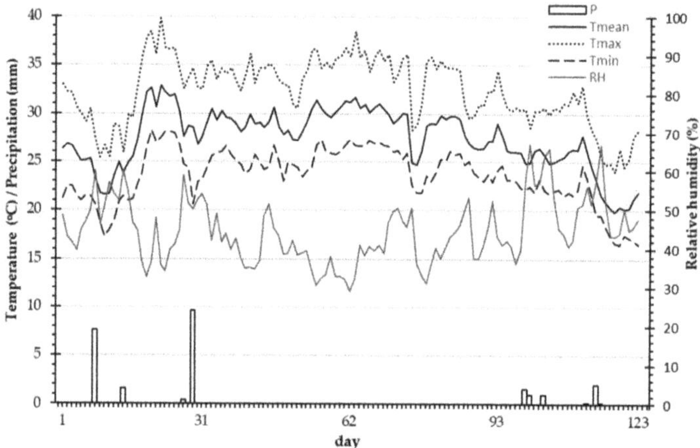

Figure 1. Diurnal mean, maximum and minimum air temperatures, precipitation and relative humidity during the simulated extensive green roof experiment on the main building of the Agricultural University of Athens from 31 May (day 1) to 30 September (day 123) in 2016. P: precipitation (mm); T_{mean}, T_{max} and T_{min}: mean, maximum and minimum temperature (°C), respectively; RH: relative humidity (%) [35].

2.6. Experimental Design and Statistical Analysis

The experiment followed a completely randomized design with four lavender species and two irrigation treatments, with six replicates per species and irrigation treatment combination. A two-way analysis of variance of the experimental data was performed using SPSS Statistical Software v. 17.0 (SPSS Inc., Chicago, U.S.A.) and treatment means were compared using Tukey HSD test at a probability level $p < 0.05$.

3. Results and Discussion

3.1. Physical-Hydraulic Properties of Substrate

Particle size distribution affects the aeration and water retention properties of substrates [37,38]. The particle size distribution of the substrate used is presented in Table 1.

Table 1. Particle size distribution of S_{15}:Pum_{70}:C_{15} substrate (subscripts show volumetric proportions of S:soil, Pum: pumice and C: grape marc compost).

Particle Size (mm)	Particle Size Distribution (% by wt)
>10	0.00
10–8	0.34
8–4	13.55
4–2	23.16
2–1	12.49
1–0.5	8.98
0.5–0.25	12.25
0.25–0.106	23.34
0.106–0.053	4.54
<0.053	1.35

Knowledge of both basic hydraulic properties of substrates, $\theta(H)$ and $K(\theta)$ is essential for irrigation management [39–41]. The measured and predicted water retention data of the substrate used are presented in Figure 2. As shown, there was a very good agreement between the experimental and predicted values of the water retention curve, indicating that the Mualem-van Genuchten model fitting parameters α, n and θ_r provide an adequate description of $\theta(H)$ with a high value of the coefficient of determination R^2 (0.9977).

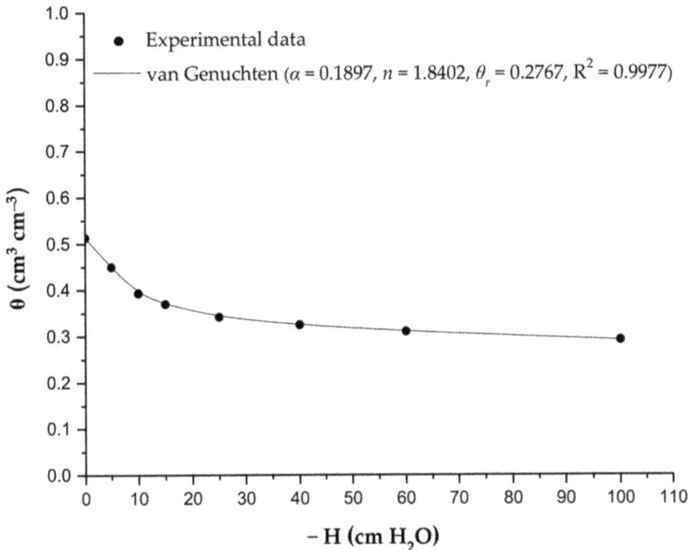

Figure 2. Experimental water retention data (symbol) and predicted curve (line) obtained by the Mualem-van Genuchten model using the RETC program for the substrate S$_{15}$:Pum$_{70}$:C$_{15}$ (subscripts show volumetric proportions of S:soil, Pum: pumice and C: grape marc compost). Values are the means of three replicates ($n = 3$).

The hydraulic characteristics derived from the water retention curve provide important information concerning plant growth and irrigation management. The main substrate hydraulic characteristics are presented in Table 2. Specifically, the total porosity (water content at 0 cm pressure head), the water content at −50 and −100 cm, as well as the easily available water (the amount of water released between −10 and −50 cm) and the air-filled porosity at −50 cm are given. Also, the measured value of hydraulic conductivity at saturation is presented.

Table 2. Hydraulic characteristics of the substrate S$_{15}$:Pum$_{70}$:C$_{15}$ (subscripts show volumetric proportions of S:soil, Pum: pumice and C: grape marc compost).

Total Porosity [1] (cm^3 cm^{-3})	Water Content at −50 cm (cm^3 cm^{-3})	Water Content at −100 cm (cm^3 cm^{-3})	Easily Available Water (EAW) [2] (cm^3 cm^{-3})	Air-Filled Porosity at −50 cm (cm^3 cm^{-3})	Ks [3] (cm min^{-1})
0.513	0.3120	0.2920	0.081	0.201	0.547

[1] water content at 0 cm pressure head (saturation); [2] the amount of water released between pressure heads of −10 and −50 cm; [3] the value of hydraulic conductivity at saturation.

In the high irrigation treatment, with the aim of retaining the substrate water content in the easily available water range, when the FDR reading reached approximately 0.31 cm^3 cm^{-3} (water content at −50 cm), plants were irrigated with the corresponding amount of water providing 100% EAW, i.e., 8.1 mm H$_2$O or 1.95 L H$_2$O. On the other hand, in the low irrigation treatment, with the aim of stressing plants, when the FDR reading reached approximately 0.29 cm^3 cm^{-3} (water content at −100 cm), plants were irrigated with half of the amount of the high irrigation treatment, i.e., 4.05 mm H_2O or 0.975 L H_2O, raising the substrate water content to approximately 0.33 cm^3 cm^{-3} (water content at −30 cm) and within the EAW range.

The unsaturated hydraulic conductivity values provide information of fundamental importance, because the rate of evapotranspiration is directly correlated to hydraulic conductivity, i.e., the water flow rate of the substrate has the ability to replace the water loss caused by evapotranspiration. In Figure 3, the measured value of hydraulic conductivity at saturation, as well as the predicted values of unsaturated hydraulic conductivity obtained by the Mualem–van Genuchten model within the range of water content between 0.513 and 0.292 cm^3 cm^{-3} (at corresponding pressure heads between 0 and −100 cm, respectively) are presented. As shown, between two successive irrigations, a sharp decrease of the unsaturated hydraulic conductivity was observed within this range. Similar results have been reported by other reasearchers on growth substrates used for plant production [39,42]. Londra [39] found a decrease of five to six orders of magnitude in unsaturated hydraulic conductivity for peat and both mixtures of peat-perlite and coir-perlite, respectively, for a pressure heads range from 0 to −70 cm. Also, Da Silva et al. [42] reported a decrease of three orders of magnitude for peat for a range from 0 to −25 cm.

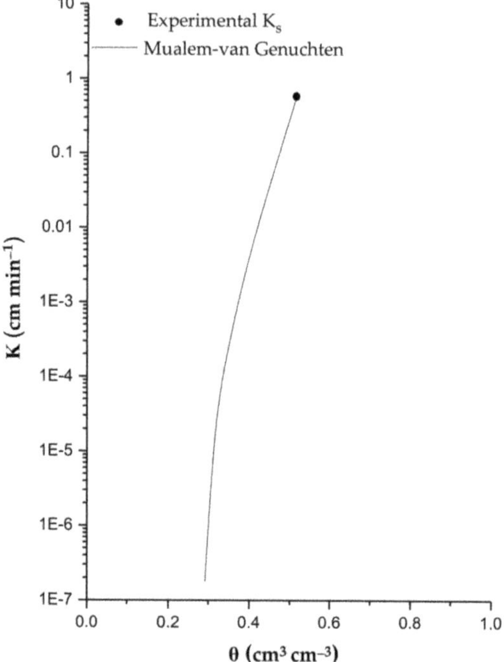

Figure 3. Experimental hydraulic conductivity at saturation (symbol) and predicted relationship between unsaturated hydraulic conductivity and water content (line) obtained by the Mualem-van Genuchten model using the RETC program for the substrate S_{15}:Pum_{70}:C_{15} (subscripts show volumetric proportions of S:soil, Pum: pumice and C: grape marc compost).

During plant growth in this study, the pressure heads varied from −10 to −50 cm between two successive irrigations in the high irrigation treatment (100% EAW) and from −30 to −100 cm in the low irrigation treatment. In the case of the high irrigation treatment, hydraulic conductivity decreased by approximately two orders of magnitude (ranged from 4.06×10^{-3} to 1.07×10^{-5} cm min^{-1}). On the other hand, in the case of the low irrigation treatment, hydraulic conductivity decreased by approximately two and a half orders of magnitude, ranging from 8.16×10^{-5} to 1.78×10^{-7} (Table 3). However, it is worth noting that the K values observed with the low irrigation treatment were much lower than those observed with the high irrigation treatment, confirming the presence of plant water stress.

Table 3. Measured values of water content (θ) and hydraulic conductivity (K) at saturation ($H = 0$ cm) and predicted ones obtained by the Mualem–van Genuchten model at water pressure heads $H = -10$, -30, and -50 cm for the substrate S_{15}:Pum_{70}:C_{15} (subscripts show volumetric proportions of S: soil, Pum: pumice and C: grape marc compost).

H (cm)	θ (cm^3 cm^{-3})	K (cm min^{-1})
0	0.513	0.547
−10	0.393	4.06×10^{-3}
−30	0.330	8.16×10^{-5}
−50	0.312	1.07×10^{-5}
−100	0.292	1.78×10^{-7}

Nevertheless, it should be noted that in some cases, the predicted $K(\theta)$ values using the water retention curve data and the saturated hydraulic conductivity may deviate significantly from the actual $K(\theta)$ values [43–47].

3.2. Symptoms Induced by Water Stress

Schroll et al. [48] mention that it is possible for shrubs that are drought tolerant in their natural habitat to be unable to use their drought tolerance mechanisms properly in the shallow, nonnative soil of an extensive green roof system. The best indicator of the effect of drought is the visible symptoms of induced damage in plants (such as leaf chlorosis, browning, and necrosis) that affects the landscape visual quality [49,50]. With the exception of *L. stoechas*, all species appeared to be healthy and demonstrated no visual signs such as chlorosis or necrosis. *L. stoechas* started to demonstrate mild leaf rolling and chlorosis of the leaves in the low irrigated water treatment approximately two months after the start of the experiment (day 62), followed one month later by *L. stoechas* irrigated with the high water treatment (day 93) (data not shown). At the end of the experiment (day 123), only *L. stoechas* demonstrated moderate leaf rolling and chlorosis of the leaves in plants irrigated with the high water treatment and severe leaf rolling and chlorosis in plants irrigated with the low water treatment (Table 4). Concerning the control (high water treatment), leaf rolling is a drought response [48]. Plants watered with the low irrigation treatment showed more intense symptoms of chlorosis compared to the corresponding plants watered with the high irrigation treatment, suggesting that the smaller amount of irrigation contributed to increasing the intensity of leaf roll and chlorosis in the plant leaves. Similar research in an extensive green roof found that *Cistus creticus* spp. *creticus* under low irrigation demonstrated brown leaves that dropped and left the branch-ends bare, which was not aesthetically-pleasing and additionally created a fire hazard due to the presence of dried leaves [48]. The presence of water stress in the low irrigation treatment was determined (see Section 3.1).

Table 4. Assessment of visual symptoms of lavender plants under different irrigation treatments at the end of the experiment (day 123), based on a 6-point scale (0–5). Differences between means shown with different letters (Tukey HSD, $p < 0.05$).

Species	Irrigation Treatment	
	High	Low
L. angustifolia	5 a	5 a
L. stoechas	3 b	2 c
L. dentata var. *candicans*	5 a	5 a
L. dentata var. *dentata*	5 a	5 a

where 0: plant mortality, 1: very severe leaf rolling and chlorosis >75%, 2: severe leaf rolling and chlorosis 50–75% approximately, 3: moderate leaf rolling and chlorosis 25–50% approximately, 4: mild leaf rolling and chlorosis <25%, 5: no leaf injury.

Furthermore, green roofs, and particularly extensive green roof systems, are characterized by the additive effect of both water deficits (water stress) and high air and substrate temperatures (heat stress) [51,52]. During warm periods, the relationship between air temperature and water in a substrate of an extensive green roof strongly influences plant growth. High substrate temperatures can limit root nutrient and water uptake and transport to leaves [53–56]. Also, the water in the substrate is susceptible to rapid evaporation [9]. In the current study, for all species and both irrigation treatments, the substrate cover from the vegetation within the surface area of each simulated extensive green roof system container was not complete (data not shown), and the meteorological data confirmed the presence of high air temperatures and moderate relative humidity (Figure 1). Further research considering substrate temperature (surface and inside) in relation to vegetation cover is necessary to study the effect of water stress on lavender species in more detail.

3.3. Plant Growth

Two-way ANOVA for data concerning the various plant growth biometrics measured throughout the duration of the experiment showed no significant interactions of the main experimental factors, i.e., among different lavender species and irrigation treatments. On the other hand, differences were shown within the experimental factors. Irrespective of lavender species, the percentage increase in height, shoot canopy diameter and growth index of plants watered with the low irrigation treatment showed smaller corresponding values than the plants watered with the high irrigation treatment ($p < 0.05$) (Table 5). Many wild plant species ceased growth due to adverse environmental conditions [57]. More specifically, during water stress, plants reduced their water requirements for the maintenance of high biomass by limiting their growth [58]. The greatest decrease in percentage increase was shown for the shoot canopy diameter. More specific, shoot canopy diameter percentage increase was reduced by 53% in plants watered with the low irrigation treatment, as opposed to plants watered with the high irrigation treatment (Table 5). Growth index percentage increase was reduced by 40% in plants watered with the low irrigation treatment, as opposed to plants watered with the high irrigation treatment. Finally, plant height percentage increase was reduced by 29% in plants watered with the low irrigation treatment, as opposed to plants watered with the high irrigation treatment. Despite the reduced percentage increase, all lavender species with the exception of *L. stoechas*, had a "healthy" appearance (i.e., without signs of leaf roll, chlorosis or necrosis). Therefore, the lavender species that showed satisfactory growth (*L. dentata* var. *candicans*, *L. dentata* var. *dentata* and *L. angustifolia*) when watered with the low irrigation treatment should be considered in the design of extensive green roof systems to create aesthetically-pleasing green roofs for urban agriculture under conditions of water stress. Note that the water content of the substrate in the low irrigation treatment remained within the substrate's available water range (see *3.2. Symptoms Induced by Water Stress*), and could contribute to conserving water resources without affecting the appearance of *L. dentata* var. *candicans*, *L. dentata* var. *dentata* and *L. angustifolia*. Further research is necessary to determine the effect of low irrigation defined by the substrate hydraulic properties of other ornamental plant species which could be grown on extensive green roofs.

Table 5. Interspecies differences and the effect of different irrigation amounts in the percentage increase in plant height (H), shoot canopy diameter (D) and growth index (GI) (n = 12, $p < 0.05$) of lavender species. Differences between means ± S.E. shown in columns with different letters (Tukey HSD, $p < 0.05$).

Species		Percentage Increase (%)		
		H	**D**	**GI**
L. dentata var. *candicans*		157 ± 7.305 a	335 ± 3.554 a	235 ± 3.669 a
L. dentata var. *dentata*		79 ± 7.305 b	277 ± 3.554 b	173 ± 3.669 b
L. angustifolia		48 ± 7.305 c	178 ± 3.554 c	107 ± 3.669 c
L. stoechas		42 ± 7.305 c	53 ± 3.554 d	38 ± 3.669 d
Irrigation treatment				
high		96 ± 5.165 a	237 ± 2.513 a	158 ± 2.595 a
low		67 ± 5.165 b	184 ± 2.513 b	118 ± 2.595 b
Interaction (species × irrigation treatment)				
L. dentata var. *candicans*	× high	179 ± 10.331	358 ± 5.025	256 ± 5.189
	× low	135 ± 10.331	313 ± 5.025	214 ± 5.189
L. dentata var. *dentata*	× high	94 ± 10.331	299 ± 5.025	191 ± 5.189
	× low	64 ± 10.331	254 ± 5.025	154 ± 5.189
L. angustifolia	× high	60 ± 10.331	213 ± 5.025	131 ± 5.189
	× low	36 ± 10.331	143 ± 5.025	84 ± 5.189
L. stoechas	× high	51 ± 10.331	78 ± 5.025	56 ± 5.189
	× low	32 ± 10.331	27 ± 5.025	21 ± 5.189
$F_{species}$/sig.		*	*	*
$F_{irrigation}$/sig.		*	*	*
$F_{interaction}$/sig.		ns	ns	ns

ns: nonsignificant; * denotes significant differences between means at $p < 0.05$, shown with different letters within columns.

Differences in the percentage increase among lavender species are likely due to interspecies variations. Overall, among the different lavender species, *L. dentata* var. *candicans* showed the greatest increase in plant height, shoot canopy diameter and growth index, followed in descending order by *L. dentata* var. *dentata*, *L. angustifolia* and *L. stoechas* with the lowest value ($p < 0.05$) (Table 5). However, *L. stoechas* showed signs of stress even under the high irrigation treatment due to the additional stress induced under the extensive green roof system (see Section 3.2); therefore is not recommended that it be used in extensive green roofs. Interspecies differences in growth provide opportunities for combining the other three lavender species in various ways and creating aesthetically-pleasing planting schemes.

On day 123, the results obtained for the shoot dry weights were consistent with the results discussed above, due to interspecies differences ($p < 0.05$). Throughout the experiment, the shoot dry weights of *L. dentata* var. *candicans* showed the greatest value, followed in descending order by *L. dentata* var. *dentata*, *L. angustifolia* and *L. stoechas* with the lowest value ($p < 0.05$) (Table 6).

Table 6. Interspecies differences and the effect of different irrigation amounts in the dry weight (DW) and stomatal conductance (mmol m^{-2} s^{-1}) (n = 12, $p < 0.05$) of lavender species. Differences between means ± S.E. shown in columns with different letters (Tukey HSD, $p < 0.05$).

Species		Shoot Dry Weight (g)	Stomatal Conductance (mmol m^{-2} s^{-1})
L. dentata var. *candicans*		188 ± 3.165 a	44 ± 1.123 c
L. dentata var. *dentata*		126 ± 3.165 b	47 ± 1.123 bc
L. angustifolia		76 ± 3.165 c	64 ± 1.123 b
L. stoechas		35 ± 3.165 d	50 ± 1.123 a
Irrigation Treatment			
high		120 ± 2.238 a	54 ± 0.794 a
low		92 ± 2.238 b	49 ± 0.794 b
Interaction (species × irrigation treatment)			
L. dentata var. *candicans*	× high	205 ± 4.477	49 ± 1.589
	× low	171 ± 4.477	45 ± 1.589
L. dentata var. *dentata*	× high	141 ± 4.477	45 ± 1.589
	× low	111 ± 4.477	43 ± 1.589
L. angustifolia	× high	86 ± 4.477	68 ± 1.589
	× low	65 ± 4.477	60 ± 1.589
L. stoechas	× high	49 ± 4.477	53 ± 1.589
	× low	21 ± 4.477	48 ± 1.589
$F_{species}$/sig.		*	*
$F_{irrigation}$/sig.		*	*
$F_{interaction}$/sig.		ns	ns

ns: nonsignificant; * denotes significant differences between means at $p <0.05$, shown with different letters within columns.

Drought avoidance physiological processes [10,20] are expressed by species in various ways, such as stomatal regulation [10]. Stomatal conductance for both species and irrigation treatments on day 123 was significant ($p < 0.05$). Among lavender species, *L. dentata* var. *candicans* showed the least stomatal conductance; this was not significantly different from the stomatal conductance of *L. dentata* var. *dentata*, followed by *L. stoechas* and *L. angustifolia*, which had the largest stomatal conductance ($p < 0.05$) (Table 6). Additionally, all lavender species showed smaller stomatal conductance values under the low irrigation treatment ($p < 0.05$) (Table 6), suggesting that the plants were being subjected to stress. As mentioned, the presence of water stress was determined (see Section 3.2). However, the stomatal conductance values in all species watered with the high irrigation treatment ($\cong$ 49–68 mmol m^{-2} s^{-1}) suggests that all species had undergone additional stress, possibly due to the additive effect of the adverse environmental conditions on the green roof (mainly by temperature) and potential root vulnerability to high substrate temperatures [18]. Substrate temperature in relation to air temperature was not studied in the present study; however, plants were exposed to high temperatures during the summer, i.e., ranging between 32.7–34.3 °C (Figure 1), and therefore, in accordance with the findings of Vestrella et al. [19], substrate temperatures may have risen by 6 °C, reaching 38.7–40.3 °C or even higher if the absolute maximum daily temperatures were considered (often above 35 °C; see Figure 1).

Low stomatal conductance values in all lavender species also suggest the presence of a drought defense strategy. Our results agree with the findings of Sendo et al. [59] that on hot summer days, in an extensive green roof system with plants not being subjected to water stress, the drought-tolerant species *Fragaria* × *ananassa*, *Thymus serphyllum*, *Evolvulus pilosus*, *Ophiopogon japonicus*, *Vinca major* and *Hedera helix* had significantly lower stomatal conductance than the nondrought tolerant species *Pelargonium* × *hortorum*, *Verbena* × *hybrida* and *Petunia* × *hybrida*. In northeast Italy (Trieste), *Salvia officinalis* grown in 14 cm deep substrate of an extensive green roof system that received natural precipitation and irrigated only during prolonged drought periods showed a stomatal conductance value of 15.1 mmol m^{-2} s^{-1} and 83% desiccation of the shoots in August that recovered by 40% after autumn rains [60]. However, *L. stoechas*, with the least stomatal conductance ($p < 0.05$) (Table 6), had symptoms of chlorosis in the leaves. The effect of water stress on the growth of the lavender species is not straightforward due to the additive effect of air and substrate temperature mentioned above. Based on the findings of Huang et al. [53] that high substrate temperatures limit root uptake, and the findings of Theodosiou [56] that plant dimensions (height and shoot canopy diameter) can reduce substrate temperature through shading, it seems that the smaller percentage increase of *L. stoechas* in relation to the other species may have led to the occurrence of chlorosis in the leaves, i.e., *L. stoechas* created less shade on the substrate surface, causing greater substrate temperature and reduced nutrient uptake. Further research is necessary to determine the effect of irrigation in relation to both air and substrate temperature. However, potential carry-over effects from year to year due to water stress need to be studied in long-lived species, such as shrubs, as the induced stress may determine plant physiological and molecular changes [61].

Throughout the experiment, the biometrics (i.e., height, shoot canopy diameter and growth index) of plants watered with the high irrigation treatment showed greater values than plants watered with the low irrigation treatment ($p < 0.05$) (Figure 4). Similarly, in an extensive green roof, *Cistus creticus* spp. *creticus* showed reduced growth index when irrigated with low amount of water compared to nonwater stress irrigated plants [48]. With the exception of *L. stoechas*, despite the smaller biometric values, the other lavender species did not demonstrate visible symptoms of induced damage by water stress. Although the low irrigation treatment produced overall smaller plants, if necessary, it could contribute to conserving water resources without affecting the appearance of *L. dentata* var. *candicans*, *L. dentata* var. *dentata* and *L. angustifolia*. As mentioned, additional research to determine the effect of low irrigation defined by the substrate hydraulic properties on the growth of other ornamental plant species of extensive green roofs is recommended.

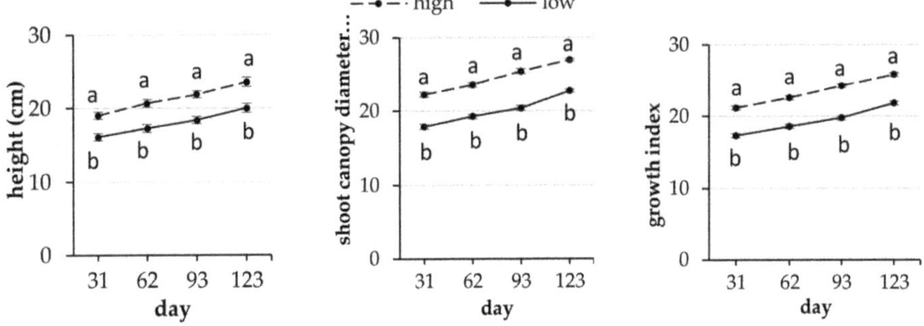

Figure 4. The effect of different irrigation amounts (high and low) on the biometrics (height, shoot canopy diameter and growth index) of *Lavandula* plants irrespective of species (n = 24, $p < 0.05$). Differences between means ± S.E. shown with different letters (Tukey HSD, $p < 0.05$) for each individual biometric variable.

Significant differences ($p < 0.05$) were also shown among different lavender species in plant height, shoot canopy diameter and growth index. In the current study, the differences in plant height among the lavender species were likely due to interspecies variations. *L. dentata* var. *candicans* showed the greatest growth in height throughout the experiment, followed by *L. dentata* var. *dentata* ($p < 0.05$). The heights of both *L. stoechas* and *L. angustifolia* were similar but smaller than the corresponding heights of the other two lavender species ($p < 0.05$) (Figure 5). The differences in height among the lavender species that showed satisfactory growth (*L. dentata* var. *candicans*, *L. dentata* var. *dentata* and *L. angustifolia*) when watered with the low irrigation treatment should be considered in the design of extensive green roof systems to create aesthetically-pleasing green roofs for urban agriculture under conditions of water stress.

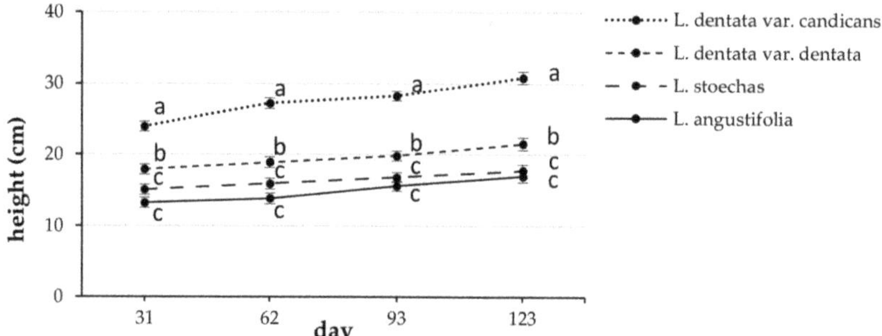

Figure 5. *Lavandula* interspecies differences in plant height (n = 12, $p < 0.05$). Differences between means ± S.E. shown in columns with different letters (Tukey HSD, $p < 0.05$).

With regards to shoot canopy diameter throughout the experiment, *L. dentata* var. *candicans* showed the greatest value, followed in descending order by *L. dentata* var. *dentata*, *L. stoechas* and *L. angustifolia* with the lowest value ($p < 0.05$) (Figure 6). These differences were due to interspecies variations. Shoot canopy diameter is an important determinant of plant success on green roofs, especially in extensive green roofs, as it influences vegetation cover [62]; vegetation cover shades the substrate surface, and hence, reduces substrate evaporation rates [18]. Therefore, the greater shoot canopy diameter of both *L. dentata* var. *candicans* and *L. dentata* var. *dentata* compared to the other lavender species suggests that they are more suitable for use in extensive green roof systems in comparison to the other lavender species studied. However, between the other two lavender species, only *L. angustifolia* showed satisfactory growth, possibly due to its dense foliage or drought tolerance mechanism [48]. As such, it is recommended for use in extensive green roof systems.

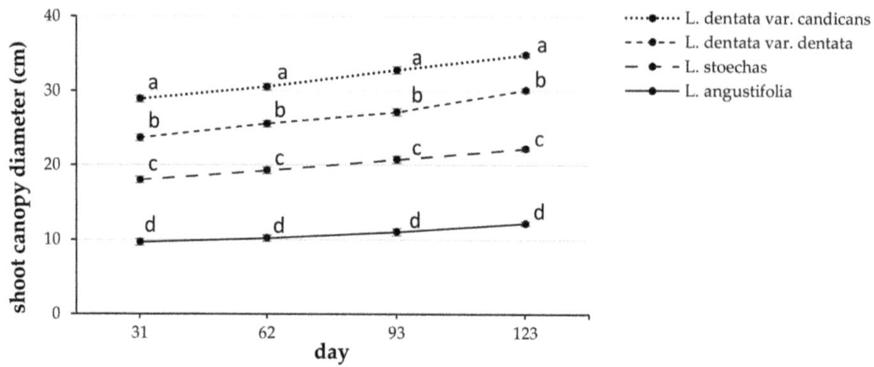

Figure 6. *Lavandula* interspecies differences in shoot canopy diameter (n = 12, $p < 0.05$). Differences between means ± S.E. shown in columns with different letters (Tukey HSD, $p < 0.05$).

Similarly, regarding the growth index throughout the experiment, *L. dentata* var. *candicans* showed the greatest value, followed in descending order by *L. dentata* var. *dentata*, *L. stoechas* and *L. angustifolia* with the lowest value ($p < 0.05$) (Figure 7). The growth index results were consistent with both the plant height and shoot canopy diameter results discussed above, and are due to interspecies differences ($p < 0.05$).

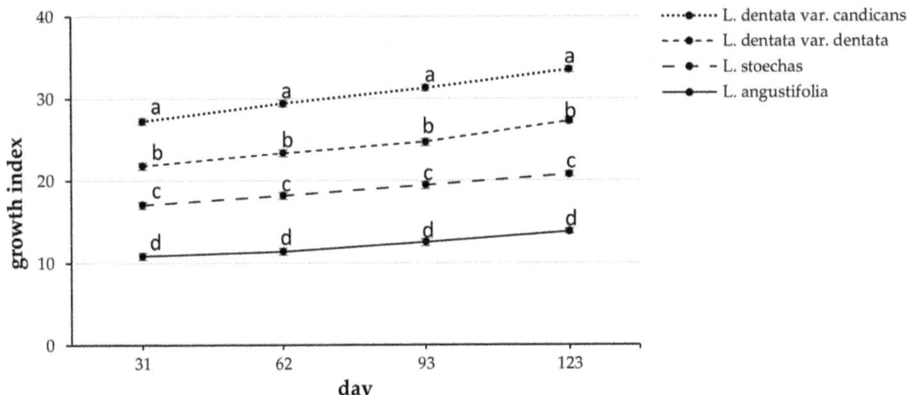

Figure 7. *Lavandula* interspecies differences in growth index (n = 12, $p < 0.05$). Differences between means ± S.E. shown in columns with different letters (Tukey HSD, $p < 0.05$).

4. Conclusions

The hydraulic properties (e.g. water retention curve, hydraulic conductivity) of substrates affect water availability and provide important information for irrigation management. In the current study, two amounts of water irrigation treatments (high and low) were applied within an available water range of an extensive green roof substrate based on its hydraulic properties. In general, plant growth was reduced in the low irrigation treatment. Among the various lavender species studied, *L. dentata* var *candicans* showed the greatest growth, while *L. angustifolia* showed the least. Overall, plant growth due to interspecies variation was as follows, in descending order: *L. dentata* var *candicans*, *L. dentata* var *dentata*, *L. stoechas* and *L. angustifolia*. All lavender species showed low stomatal conductance values, suggesting the presence of a drought defense strategy. *L. dentata* var. *candicans* showed the lowest stomatal conducatance value, similar to those of *L. dentata* var. *dentata* and followed in ascending order by *L. stoechas* and *L. angustifolia*, with the greatest stomatal conductance.

On the other hand, with the exception of *L. stoechas*, the appearance of all species studied was satisfactory. Therefore, the use of *L. stoechas* is not proposed, as the visual quality of the plant was reduced due to leaf roll and chlorosis induced by stress. Despite the reduced growth of lavender species watered with the low irrigation treatment, the satisfactory appearance (i.e., lack of damage induced symptoms) of the plants supports the use of *L. dentata* var *candicans*, *L. dentata* var *dentata*, and *L. angustifolia* under low irrigation. Further study on the effect of low irrigation within the substrate available water range determined by the substrate hydraulic properties on the growth of other ornamental plant species on extensive green roofs is recommended.

The differences in plant characteristics and size among *L. dentata* var *candicans*, *L. dentata* var *dentata*, and *L. angustifolia* can be considered in the design of extensive green roof systems including amphitheatrical planting schemes to create aesthetically-pleasing green roofs for urban agriculture. The larger *L. dentata* var *candicans* and *L. dentata* var *dentata* varieties are recommended for extensive green roofs, as they provide greater vegetation cover in substrates, potentially reducing the effect of high substrate temperatures due to shading. Further research is recommended to determine the effect of irrigation in relation to drought tolerance mechanisms, as well as both air and substrate temperatures in extensive green roof systems.

Author Contributions: Conceptualization, A.T.P.; methodology, A.T.P. and P.A.L.; formal analysis, A.T.P., P.T., P.A.L. and A.P.K.; investigation, P.T. and A.P.; writing—original draft preparation, A.P.; writing—review and editing, A.T.P., P.A.L. and A.P.K.; supervision, A.T.P. and P.A.L.; project administration, A.P. and P.T. All authors have read and agreed to the published version of the manuscript.

Acknowledgments: We would like to thank Kalantzis Plants for the free supply of lavender plants.

Conflicts of Interest: The authors declare no conflict of interest.

References

1. Azunre, G.A.; Amponsah, O.; Peprah, C.; Takyi, S.A.; Braimah, I. A review of the role of urban agriculture in the sustainable city discourse. *Cities* **2019**, *93*, 104–119. [CrossRef]
2. Mougeot, L.J.A. *Urban. Agriculture: Definition, Presence, Potentials and Risks, and Policy Challenges*; Cities Feeding People Series Report 31; International Development Research Centre (IDRC): Ottawa, ON, Canada, 2000; 62p.
3. Marques-Perez, I.; del Rio, B.S.G. Identifying Functionality of Peri-Urban Agricultural Systems: A Case Study. In *Urban Agriculture*; Samer, M., Ed.; InTech: London, UK, 2016; pp. 61–88.
4. Artmann, M.; Sartison, K. The Role of Urban Agriculture as a Nature-Based Solution: A Review for Developing a Systemic Assessment Framework. *Sustainability* **2018**, *10*, 1–32. [CrossRef]
5. Dunnett, N.; Kingsbury, N. *Planting Green Roofs and Living Walls*; Timber Press: Portland, OR, USA, 2004; 254p.
6. Dubbeling, M.; Orsini, F.; Gianquinto, G. The Status and Challenges of Rooftop Agriculture. In *Rooftop Urban Agriculture*; Orsini, F., Dubbeling, M., de Zeeuw, H., Gianquinto, G., Eds.; Springer International Publishing: Cham, Switzerland, 2017; pp. 3–8.
7. Besir, A.B.; Cuce, E. Green roofs and facades: A comprehensive review. *Renew. Sustain. Energy Rev.* **2018**, *82*, 915–939. [CrossRef]
8. Paraskevopoulou, A.; Mitsios, I.; Fragkakis, I.; Nektarios, P.; Ntoulas, N.; Londra, P.; Papafotiou, M. The growth of *Arthrocnemum macrostachyum* and *Halimione portulacoides* in an extensive green roof system under two watering regimes. *Agric. Agric. Proc.* **2015**, *4*, 242–249. [CrossRef]
9. Savi, T.; Boldrin, D.; Marin, M.; Love, V.L.; Andri, S.; Trtiah, M.; Nardini, A. Does shallow substrate improve water status of plants growing on green roofs? Testing the paradox in two sub-Mediterranean shrubs. *Ecol. Eng.* **2015**, *84*, 292–300. [CrossRef]
10. Du, P.; Arndt, S.K.; Farrell, C. Is plant survival on green roofs related to their drought response, water use or climate or origin? *Sci. Total Environ.* **2019**, *667*, 25–32. [CrossRef]
11. Papafotiou, M.; Pergalioti, N.; Tassoula, L. Growth of Native Aromatic Xerophytes in an Extensive Mediterranean Green Roof as Affected by Substrate Type and Depth and Irrigation Frequency. *HortScience* **2013**, *48*, 1327–1333. [CrossRef]
12. Papafotiou, M.; Pergalioti, N.; Massas, I.; Kargas, G. Effect of Substrate Type and Depth and the Irrigation Frequency on Growth of Semiwoody Mediterranean Species in Green Roofs. *Acta Hortic.* **2013**, *990*, 481–486. [CrossRef]
13. Tassoula, L.; Papafotiou, M.; Liakopoulos, G.; Kargas, G. Growth of the Native Xerophyte *Convolvulus cneorum* L. on an Extensive Mediterranean Green Roof under Different Substrate Types and Irrigation Regimens. *HortScience* **2015**, *50*, 1118–1124. [CrossRef]
14. Papafotiou, M.; Tassoula, L.; Liakopoulos, G.; Kargas, G. Effect of substrate type and irrigation frequency on growth of Mediterranean xerophytes on green roofs. *Acta Hortic.* **2016**, *1108*, 309–315. [CrossRef]
15. Papafotiou, M.; Tassoula, L.; Kefalopoulou, R. Effect of substrate type and irrigation frequency on growth of *Pallenis maritima* on an urban extensive green roof at the semi-arid Mediterranean region. *Acta Hortic.* **2017**, *1189*, 275–278. [CrossRef]
16. Tassoula, L.; Papafotiou, M.; Fouskaki, M. Growth of the halophyte *Atriplex halimus* on a green roof at the semi-arid Mediterranean region as affected by substrate type and irrigation regime. *Acta Hortic.* **2017**, *1189*, 287–290. [CrossRef]
17. Baltzoi, P.; Fotia, K.; Kyrkas, D.; Nikolaou, K.; Paraskevopoulou, A.T.; Accogli, A.R.; Karras, G. Low water–demand plants for landscaping and agricultural cultivations—A review regarding local species of Epirus/Greece and Apulia/Italy. *Agric. Agric. Proc.* **2015**, *4*, 250–260. [CrossRef]
18. Wolf, D.; Lundholm, J.T. Water uptake in green roof microcosms: Effects of plant species and water availability. *Ecol. Eng.* **2008**, *33*, 179–186. [CrossRef]
19. Vestrella, A.; Biel, C.; Savè, A.; Bartoli, F. Mediterranean Green Roof Simulation in Caldes de Montbui (Barcelona): Thermal and Hydrological Performance Test of *Frankenia laevis* L., *Dymondia margaretae* Compton and *Iris lutescens* Lam. *Appl. Sci.* **2018**, *8*, 2497. [CrossRef]

20. Bartlett, M.K.; Klein, T.; Jansen, S.; Choat, B.; Sack, L. The correlations and sequence of plant stomatal, hydraulic, and wilting responses to drought. *PNAS* **2016**, *113*, 13098–13103. [CrossRef] [PubMed]

21. Mansfield, T.J.; Atkinson, C.J. Stomatal behavior in water stressed plants. In *Stress Responses in Plants: Adaptation and Acclimation Mechanisms*; Alscher, R.G., Cumming, J.R., Eds.; Wiley-Liss: New York, NY, USA, 1990; pp. 241–264.

22. Arve, L.E.; Torre, S.; Olsen, J.E.; Tanino, K.K. Stomatal Responses to Drought Stress and Air Humidity. In *Abiotic Stress in Plants—Mechanisms and Adaptations*; Arun, K.S., Venkateswarlu, B., Eds.; Intech: Rijeka, Croatia, 2011; pp. 279–280.

23. Pirasteh-Anosheh, H.; Saed-Moucheshi, A.; Pakniyat, H.; Pessarakli, M. Stomatal responses to drought stress. In *Water Stress and Crop. Plants: A Sustainable Approach*, 1st ed.; Ahmad, P., Ed.; John Wiley & Sons, Ltd.: Chichester, UK, 2016; Volume 1, pp. 24–40.

24. Torres-Ruiz, J.M.; Diaz-Espejo, A.; Morales-Sillero, A.; Martín- Palomo, M.J.; Mayr, S.; Beikircher, B.; Fernandez, J.E. Shoot hydraulic characteristics, plant water status and stomatal response in olive trees under different soil water conditions. *Plant. Soil* **2013**, *373*, 77–87. [CrossRef]

25. Casson, S.A.; Hetherington, A.M. Environmental regulation of stomatal development. *Curr. Opinion Plant. Biol.* **2010**, *13*, 90–95. [CrossRef]

26. De Boodt, M.; Verdonck, O. The physical properties of the substrates in horticulture. *Acta Hortic.* **1972**, *26*, 37–44. [CrossRef]

27. The Plant List, Version 1.1. 2013. Available online: http://www.theplantlist.org/1.1/browse/A/Lamiaceae/Lavandula/#statistics (accessed on 18 March 2020).

28. Upson, T. The taxonomy of the genus *Lavandula*, L. In *Lavender The genus Lavandula*; Lis-Balchin, M., Ed.; Taylor & Francis: London, UK, 2002; pp. 2–34.

29. Lis-Balchin, M. General introduction to the genus *Lavandula*. In *Lavender The genus Lavandula*; Lis-Balchin, M., Ed.; Taylor & Francis: London, UK, 2002; p. 1.

30. Haines, W.B. Studies in the physical properties of soils. V. The hysteresis effect in capillary properties and the modes of moisture distribution associated therewith. *J. Agric. Sci.* **1930**, *20*, 97–116. [CrossRef]

31. Van Genuchten, M.T.; Leij, F.J.; Yates, S.R. *The RETC Code for Quantifying the Hydraulic Functions of Unsaturated Soils*; U.S.D.A. (U.S. Department of Agriculture, Agricultural Research Service): Riverside, CA, USA, 1991.

32. Mualem, Y. A new model for predicting the hydraulic conductivity of unsaturated porous media. *Water Resour. Res.* **1976**, *12*, 513–522. [CrossRef]

33. Van Genuchten, M.T. A closed-form equation for predicting the hydraulic conductivity of unsaturated soils. *Soil Sci. Soc. Am. J.* **1980**, *44*, 892–898. [CrossRef]

34. Klute, A.; Dirksen, C. *Methods of Soil Analysis, Part 1, Physical and Mineralogical Methods*; Klute, A., Ed.; American Society of Agronomy: Madison, WI, USA, 1986; pp. 694–700.

35. IERSD-NOA. *Monthly Bulletin*; Institute of Environmental Research and Sustainable Development of the National Observatory of Athens: Penteli, Greece, 2016; Available online: https://meteo.gr/Monthly_Bulletins.cfm (accessed on 7 January 2020).

36. W.M.O. *Guide to Meteorological Instruments and Methods of Observation*; WMO-No. 8 (updated 2017); World Meteorological Organization (W.M.O.): Geneva, Switzerland, 2014; 1166p.

37. Raviv, M.; Lieth, J.H. *Soilless Culture Theory and Practice*; Elsevier BV: London, UK, 2008; p. 587.

38. Naaz, R.; Bussières, P. Particle Sizes Related to Physical Properties of Peat-Based Substrates. *Acta Hortic.* **2011**, *893*, 971–978. [CrossRef]

39. Londra, P.A. Simultaneous determination of water retention curve and unsaturated hydraulic conductivity of substrates using a steady-state laboratory method. *HortScience* **2010**, *45*, 1106–1112. [CrossRef]

40. Londra, P.A.; Paraskevopoulou, A.T.; Psychoyou, M. Evaluation of water-air balance of various substrates on begonia growth. *HortScience* **2012**, *47*, 1153–1158. [CrossRef]

41. Londra, P.A.; Psychoyou, M.; Valiantzas, J.D. Evaluation of substrate hydraulic properties amended by urea-formaldehyde resin foam. *HortScience* **2012**, *47*, 1375–1381. [CrossRef]

42. Da Silva, F.F.; Wallach, R.; Chen, Y. Hydraulic properties of Sphagnum peat moss and tuff (scoria) and their potential effects on water availability. *Plant. Soil* **1993**, *154*, 119–126. [CrossRef]

43. Talsma, T. Prediction of hydraulic conductivity from soil water retention data. *Soil Sci.* **1985**, *140*, 184–188. [CrossRef]

44. Poulovassilis, A.; Polychronides, M.; Kerkides, P. Evaluation of various computational schemes in calculating unsaturated hydraulic conductivity. *Agric. Water Manag.* **1988**, *13*, 317–327. [CrossRef]

45. Valiantzas, J.D.; Sassalou, A. Laboratory determination of unsaturated hydraulic conductivity using a generalized-form hydraulic model. *J. Hydrol.* **1991**, *128*, 293–304. [CrossRef]

46. Kargas, G.; Londra, P.A. Effect of tillage practices on the hydraulic properties of a loamy soil. *Desalin. Water Treat.* **2015**, *54*, 2138–2146. [CrossRef]

47. Londra, P.; Kargas, G. Evaluation of hydrodynamic characteristics of porous media from one-step outflow experiments using RETC code. *J. Hydroinform.* **2018**, *20*, 699–707. [CrossRef]

48. Schroll, E.; Lambrinos, J.G.; Sandrock, D. An Evaluation of Plant Selections and Irrigation Requirements for Extensive Green Roofs in the Pacific Northwestern United States. *HortTechnology* **2011**, *21*, 314–322. [CrossRef]

49. Vahdati, N.; Tehranifar, A.; Kazemi, F. Assessing chilling and drought tolerance of different plant genea on extensive green roofs in an arid climate region in Iran. *J. Environ. Manage.* **2017**, *192*, 215–223. [CrossRef] [PubMed]

50. Nazemi Rafi, Z.; Kazemi, F.; Tehranifar, A. Effects of various irrigation regimes on water use efficiency and visual quality of some ornamental herbaceous plants in the field. *Agr. Water Manage.* **2019**, *212*, 78–87. [CrossRef]

51. Savi, T.; Dal Borgo, A.; Love, V.L.; Andri, S.; Tretiach, M.; Nardini, A. Drought versus heat: What's the major constraint on Mediterranean green roof plants? *Sci. Total Environ.* **2016**, *566–567*, 753–760. [CrossRef]

52. Kazemi, F.; Mohorko, R. Review on the roles and effects of growing media on plant performance in green roofs in world climates. *Urban. Forestry and Urban. Greening* **2017**, *23*, 13–26. [CrossRef]

53. Huang, B.; Rachmilevitch, S.; Xu, J. Root carbon and protein metabolism associated with heat tolerance. *J. Experimental Botany* **2012**, *63*, 3455–3465. [CrossRef]

54. Olivieri, F.; Di Perna, C.; D'Orazio, M.; Olivieri, L.; Neila, J. Experimental measurements and numerical model for the summer performance assessment of extensive green roofs in a Mediterranean coastal climate. *Energ. Buildings* **2013**, *63*, 1–14. [CrossRef]

55. Simmons, M.T.; Gardiner, B.; Windhager, S.; Tinsley, J. Green roofs are not created equal: The hydrologic and thermal performance of six different extensive green roofs and reflective and non-reflective roofs in a sub-tropical climate. *Urban. Ecosyst.* **2008**, *11*, 339–348. [CrossRef]

56. Theodosiou, T.G. Summer period analysis of the performance of a planted roof as a passive cooling technique. *Energ. Buildings* **2003**, *35*, 909–917. [CrossRef]

57. Chapin, F.S., III. The mineral nutrition of wild plants. *Ann. Rev. Ecol. Syst.* **1983**, *11*, 233–260. [CrossRef]

58. Schuppler, U.; He, P.H.; John, P.C.L.; Munns, R. Effects of water stress on cell division and cell-division-cycle2-like cell-cycle kinase activity in wheat leaves. *Plant. Physiol.* **1998**, *117*, 667–678. [CrossRef] [PubMed]

59. Sendo, T.; Kanechi, M.; Uno, Y.; Inagaki, N. Evaluation of Growth and Green Coverage of Ten Ornamental Species for Planting as Urban Rooftop Greening. *J. Japan. Soc. Hort. Sci.* **2010**, *79*, 69–76. [CrossRef]

60. Savi, T.; Andri, S.; Nardini, A. Impact of different green roof layering on plant water status and drought survival. *Ecol. Eng.* **2013**, *57*, 188–196. [CrossRef]

61. Mirás-Avalos, J.M.; Pérez-Sarmiento, F.; Alcobendas, R.; Alarcón, J.J.; Mounzer, O.; Nicolás, E. Using midday stem water potential for scheduling deficit irrigation in mid–late maturing peach trees under Mediterranean conditions. *Irr. Sci.* **2016**, *32*, 161–173. [CrossRef]

62. Molineux, C.J.; Fentiman, C.H.; Gange, A.C. Characterising alternative recycled waste materials for use as green roof growing media in the U.K. *Ecol. Eng.* **2009**, *35*, 1507–1513. [CrossRef]

© 2020 by the authors. Licensee MDPI, Basel, Switzerland. This article is an open access article distributed under the terms and conditions of the Creative Commons Attribution (CC BY) license (http://creativecommons.org/licenses/by/4.0/).

Article

The Effect of Salinity on the Growth of Lavender Species

Angeliki T. Paraskevopoulou [1,*], Anna Kontodaimon Karantzi [1], Georgios Liakopoulos [2], Paraskevi A. Londra [3] and Konstantinos Bertsouklis [1]

[1] Laboratory of Floriculture and Landscape Architecture, Department of Crop Science, School of Plant Sciences, Agricultural University of Athens, Iera Odos 75, 118 55 Athens, Greece; annakontodaimon9225@gmail.com (A.K.K.); kber@aua.gr (K.B.)

[2] Laboratory of Plant Physiology and Morphology, Department of Crop Science, School of Plant Sciences, Agricultural University of Athens, Iera Odos 75, 118 55 Athens, Greece; gliak@aua.gr

[3] Laboratory of Agricultural Hydraulics, Department of Natural Resources Management & Agricultural Engineering, School of Environment and Agricultural Engineering, Agricultural University of Athens, Iera Odos 75, 118 55 Athens, Greece; v.londra@aua.gr

* Correspondence: aparas@aua.gr; Tel.: +30-210-529-4551

Received: 10 February 2020; Accepted: 23 February 2020; Published: 25 February 2020

Abstract: Long term degradation of water quality from natural resources has led to the use of alternative water resources for irrigation that are saline. Saline water irrigation in floriculture for the production of nursery crops requires an understanding of plant response. The pot growth of four lavender species (*Lavandula angustifolia*, *Lavandula dentata* var. *dentata*, *Lavandula dentata* var. *candicans* and *Lavandula stoechas*) irrigated with water containing different concentrations of NaCl (0, 25, 50, 100 and 200 mM) was investigated under greenhouse conditions. Overall results of different plant growth variables were consistent, showing a significant decrease at 100 and 200 mM NaCl. All lavender species showed signs of salinity stress that included chlorosis, followed by leaf and stem necrosis at NaCl concentrations greater than 50 mM. *L. dentata* var. *dentata* showed the greatest plant growth followed in descending order by *L. dentata* var. *candicans*, *L. stoechas* and *L. angustifolia*. Despite greater growth of *L. dentata* var. *dentata*, the appearance of *L. dentata* var. *candicans* was "healthier". In areas with saline irrigation water, *L. dentata* var. *dentata* and *L. dentata* var. *candicans* are proposed for the production of lavender nursery crops.

Keywords: floriculture; *Lavandula angustifolia*; *Lavandula dentata* var. *dentata*; *Lavandula dentata* var. *candicans*; *Lavandula stoechas*; saline water; irrigation; NaCl; chlorophyll fluorescence

---

1. Introduction

In arid, semi-arid and coastal areas, natural resources for good quality water have decreased. They are often characterized by high contents of total soluble salts due to groundwater overexploitation, seawater intrusion into aquifers and increased demand for freshwater, particularly in densely populated areas [1,2]. Long term degradation of water quality has led to the use of alternative water resources for irrigation derived from water reuse and recycling that is also saline [3,4]. Irrigation with saline water affects the growth and development of many plant species, even at low concentrations [5,6]. In the ground whether in the wild, field or garden, the effect of salinity on plants is determined by various variables such as ion concentration, soil composition, proximity to the sea, altitude, evapotranspiration rate, temperature and rainfall frequency [5–7]. In many parts of the world, salinity affects agricultural production and is predicted to become more intense in future decades [8]. It is considered as one of the most important stress factors in plant growth and yield that could lead to plant death under persisting saline conditions [6,9]. Plant tolerance to salinity stress depends on the capacity of plants to exclude salt from the shoots or tolerate high leaf salt concentrations [10].

Irrigation with saline water initially creates a water deficit induced by osmotic stress and demonstrated by the reduced ability of plants to absorb water hence reduced plant growth rate [11]; the high saline concentrations cause osmotic and ionic imbalances between soil and plants, and plants exhibit signs of wilt despite the fact they have been irrigated [12,13]. Afterwards, a salt-specific or ion-excess effect of salinity is demonstrated by the salt entrance into the plant transpiration stream, causing eventual injury of transpiring leaf cells and further reduction of plant growth [11]. The high saline concentrations within the plant affect the anatomy, physiology and morphology of plant parts and particularly of leaves [4,6,14,15]. The salts absorbed by the plant are concentrated within the mature leaves, leading to leaf death over an extended time period due to the inability of leaf cells to compartmentalize salts in the vacuole; hence the salts either accumulate in the cytoplasm, inhibiting enzyme function, or accumulate in the cell walls, dehydrating the leaf cells [11]. The level of stress caused by salinity is dependent on the plant species and variety, the growth substrate and the applied method of irrigation. The more tolerant nonhalophytic species avoid the ion-excess effect. However, they may exhibit water deficits affecting cell extension and/or division. Therefore, potential reductions in photosynthesis may represent a secondary effect of reduced growth [16].

In floriculture, the use of saline water for the production of nursery crops requires an understanding of plant response to the effect of salinity through irrigation [17]. Some effects of salinity, on one hand, could be desirable such as decreased length and/or number of internodes [17] and others on the other hand could be undesirable such as chlorosis and marginal leaf necrosis. The effect of saline irrigation on floriculture has received less attention, as ornamental plants are normally irrigated with good-quality water [18,19]. In areas with limited or poor water quality resources, the cultivation of floriculture crops that can tolerate saline water irrigation can be an advantage [20]. Lavender species and varieties are popular floriculture crops. Lavender plants such as *Lavandula angustifolia*, *L. dentata*, and *L. stoechas* and their numerous cultivars are sold as ornamental plants for the garden. These species exhibit a variety of leaves and inflorescences with ornamental value and are highly aromatic due to the essential oils present in glands that cover much of the plant surface. The *Lavandula* genus includes 47 species and many varieties [21]. Some *Lavandula* species such as *Lavandula stoechas* and *Lavandula angustifolia* are found naturally growing in the Mediterranean coniferous coastal dune woodlands, coastal garrigues and sea cliffs, often exposed to sea spray [22,23]. The literature on the effect of saline irrigation on *Lavandula* species for nursery crops is limited. Potted *L. multifida* plants were able to grow in a mixture of sphagnum peat-moss and perlite when irrigated with 60 mM NaCl without significant biomass reduction [24]; however, the total plant dry weight of *L. multifida* decreased when irrigated with 100 mM and 200 mM NaCl [20,24]. Despite this, there are no comparative studies among different lavender species grown under greenhouse conditions and irrigated with different NaCl solutions.

This study examines the pot growth of four *Lavandula* species irrigated with different concentrations of saline water for nursery production to support floriculture in areas with poor water quality, using saline water for irrigation.

2. Materials and Methods

2.1. Experimental Site and Growth Conditions

Uniform in size, young (5 months old) and fully developed potted lavender plants were supplied by the nursery Kalantzis Plants (Marathonas, Greece). The pot size was 2.5 L (dimensions: 17 cm top diameter, 12.3 cm base diameter and 15 cm height). The growth substrate of the supplied potted lavender plants contained pure sphagnum peat (Base Substrate 2 medium, Klausmann-Deilmann Europe GmbH, Germany) and perlite (Perloflor, ISOCON S.A., Piraeus, Greece) in a 96:4 ratio (v/v) with pH 5.5–6.0 and EC 0.8 mS m^{-1}. Plants were placed on metal benches (dimensions: 2.5 m length, 0.85 m width and 0.80 m height) in an automated glass greenhouse of the Laboratory of Floriculture &

Landscape Architecture of the Agricultural University of Athens (lat. 37°58′57″N and long. 23°42′17″E), with average daily and night temperatures of 21.4 ± 0.311 °C and 14.3 ± 0.065 °C, respectively, and average humidity during daytime of 57.6 ± 0.705% and night-time average humidity of 84.6 ± 0.309%. Plants were acclimatized to the new growth conditions for a month and the experiment took place in late winter-early spring over 56 days (from 3 February; day 1 to 30 March 2018; day 56). All plants received the same cultivation practices (i.e., applications of fertilizer, fungicide, etc.) throughout the duration of the experiment that included the application of 2 g L^{-1} H$_2$O fertilizer 20-20-20 (Fast-Grow, Humofert S.A., Metamorfosi, Greece) and pesticide (Decis 25 EC, Bayer AG, Leverkusen, Germany) at monthly intervals.

2.2. Experimental Design and Irrigation Treatments

Four lavender species were studied: *Lavandula angustifolia*, *Lavandula dentata* var. *dentata*, *Lavandula dentata* var. *candicans* and *Lavandula stoechas*. The effect of salinity was investigated using different concentrations of NaCl solutions through irrigation that included 0 (control), 25, 50, 100 and 200 mM of NaCl. The corresponding EC levels for the irrigation water were 0.3, 3.0, 5.8, 10.6 and 20.7 dSm^{-1} and pH values were in the range 8.0–8.2 (at 25 °C).

Plants were arranged in a randomized complete block that consisted of 4 lavender species, 5 NaCl solution irrigation treatments and 6 replicates (plants) arranged in 3 blocks (metal benches) i.e., 2 plants per species and NaCl solution irrigation treatment per metal bench. The number of plants totaled 120 and the experimental surface area occupied approximately 6.5 m^2 (Figure 1).

At the start of the experiment (day 1), all plants were irrigated with the corresponding NaCl solutions to saturation and weighed half an hour later to determine the water container capacity of the substrate. Substrate water content was monitored using a handheld TDR moisture sensor (HH2, Delta-T Devise, Cambridge, UK) set at the 'organic soil' setting, appropriate for use with peat-based substrates and calibrated to the used substrate. The probes were fully inserted into the substrate with the central rod positioned 5 cm away from the plant center. Irrigation was performed manually when the TDR sensor showed a water content value of approximately 0.46 cm^3 cm^{-3}, which was determined from the substrate water retention curve at corresponding a pressure head of −50 cm (Figure 2), and with an amount of water ensuring substrate water availability within the easily available water area (Table 1). This amount of irrigation water of plants was determined with the mean accumulated daily difference in weight of six potted plants from each NaCl treatment between two consecutive irrigations that corresponded to the amount of water lost from evapotranspiration.

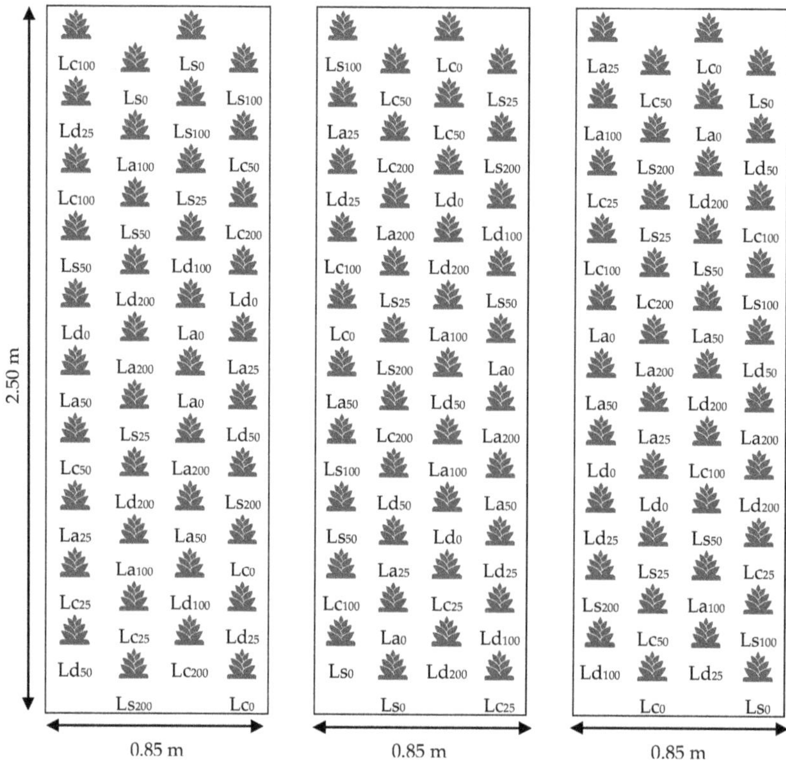

Figure 1. The layout of the experiment studying the effect of different NaCl solution irrigation treatments on the growth of 4 lavender species. Plants were arranged in a randomized complete block that consisted of 4 lavender species, 5 NaCl solution irrigation treatments (0, 25, 50, 100, 200 mM NaCl) and 6 replicates (plants) arranged in 3 blocks (metal benches; dimensions 2.50 cm length and 0.85 m width). La: *Lavandula angustifolia*, Ld: *Lavandula dentata* var. *dentata*, Lc: *Lavandula dentata* var. *candicans*, Ls: *Lavandula stoechas*, and subscripts denote applied NaCl solution irrigation treatments.

2.3. Plant Growth Variables

Measurements started one week (day 7) after irrigation with the NaCl solutions for the first time and ended 56 days later. Plant height (determined from the pot rim of the substrate surface), shoot canopy diameter (mean value of the widest width and perpendicular width), and growth index ((height + widest width + perpendicular width)/3) were measured at weekly intervals. Additionally, during flowering, for each plant, the number and length of all inflorescences that were fully open >60% as well as the corresponding peduncle length, were recorded at weekly intervals. The maximum efficiency of PSII photochemistry (Φ_{PSIIo}) of mature leaves (3 leaves per plant) was determined fortnightly (day 14, 35, 56) using a MINI-PAM Photosynthesis Yield Analyzer (Heinz Walz GmbH, Effeltrich, Germany). All measurements were performed in the morning after dark acclimation of the samples for 30 min using the saturation pulse technique. Saturation pulse (intensity circa 12,000 μmol quanta m^{-2} s^{-1}) lasted 0.8 s.

At the end of the experiment (day 56) the leaf thickness of mature leaves was determined with cross sections taken at a distance of 3 cm from the leaf base (3 leaves per plant) under a Zeiss Axiolab microscope (Carl Zeiss, Jena, Germany) using the x100 lens. Plants were harvested at the end of the experiment (day 56) and divided at soil level into shoot and root. The substrate was carefully washed off the harvested root. Following both harvested shoots and roots were separately dried in an oven at

70 °C until a constant weight was reached, and their dry weights were determined. Weekly recordings of observations for signs of salinity stress were undertaken, throughout the duration of the experiment. In plants, the onset of visual symptoms induced by salinity (leaf and stem chlorosis and necrosis) was recorded during the experiment and assessed at the end of the experiment (day 56). Visual symptoms induced by salinity were assessed on a 6 point scale from 0–5, where 0: plant mortality, 1: no leaf injury, 2: mild leaf chlorosis, 3: moderate leaf chlorosis 25–50% approximately, 4: leaf necrosis 50–75% approximately, 5: leaf necrosis >75%.

2.4. Physical-Hydraulic Properties of Substrate

A tension plate apparatus in a Haines-type assembly [25], with an air-entry value of −180 cm of a water column was employed to define the substrate water retention curve. Substrate sample sized 3 cm in height and 10.2 cm in diameter was positioned on the vibrating porous plate of a Buchner filter funnel to achieve satisfactory packing and following was subjected to gradual wetting from the bottom of the plate until saturation (for 48 h). Measurements of the water content at different pressure heads were taken to obtain the water retention curve. The retention curve was the mean of three substrate samples (n = 3).

Particle size distribution of the substrates was determined with screen analysis. Weighed substrate samples were placed in the top sieve of a column of sieves arranged from top to bottom in descending order of screen mesh size (>20.00, 16.00, 10.00, 8.00, 4.00, 2.00, 1.00, 0.50, 0.25, 0.106 and <0.053 mm) rested on a sieve shaker for 3 min at 30 shakes per minute.

2.5. Statistical Analysis

The experiment followed a randomized complete block design with two factors that constituted of four lavender species and five NaCl solution irrigation treatments. There were six replications (plants) per species and NaCl solution irrigation treatment. A two-way analysis of variance (ANOVA) was applied to test the significance of the experimental data using SPSS Statistical Software v. 17.0 (SPSS Inc., Chicago, U.S.A.), and treatment means were compared using Tukey HSD test at a probability level $p < 0.05$.

3. Results and Discussion

3.1. Physical–Hydraulic Properties of Substrate

Pressure heads of the experimental water retention curve of the substrate ranged between 0 and −100 cm providing important information concerning plant growth (Figure 2). In accordance to De Boodt and Verdonck [26], water retention in substrates with negative pressure heads greater than −100 cm decreases plant growth, while a negative pressure head that is less than −10 cm creates inadequate substrate aeration for plant growth. The main substrate hydraulic characteristics derived from the water retention curve are shown in Table 1.

Table 1. Hydraulic characteristics of the potted lavender plants' growth substrate Ps96:P4 (Ps: pure sphagnum peat and P: perlite in a 96:4 volume ratio).

Total Porosity [1] (cm³ cm⁻³)	Airspace [2] (cm³ cm⁻³)	Water Content at −10 cm (cm³ cm⁻³)	Water Content at −50 cm (cm³ cm⁻³)	Easily Available Water [3] (cm³ cm⁻³)	Water Buffering Capacity [4] (cm³ cm⁻³)
0.88	0.12	0.76	0.46	0.30	0.13

[1] water content at 0 cm pressure head (saturation); [2] air filled pores at −10 cm pressure head; [3] released amount of water between pressure heads of −10 and −50 cm; [4] released amount of water between pressure heads of −50 and −100 cm.

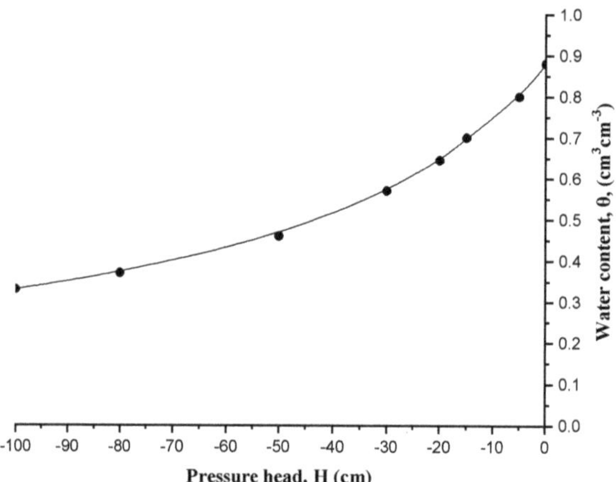

Figure 2. Water retention curve of the potted lavender plants' growth substrate Ps$_{96}$:P$_4$ (Ps: pure sphagnum peat and P: perlite in a 96:4 volume ratio).

With an aim to retain the substrate water content in the easily available water range, i.e., water content between 0.76 and 0.46 cm^3 cm^{-3} (at −10 and −50 cm pressure head, respectively) (Table 1, Figure 2), when the TDR reading reached a volumetric water content of 46%, plants were irrigated with the corresponding NaCl treatments at amounts equal to the amount lost from evapotranspiration.

Particle size distribution affects the aeration and water retention properties of substrates [27,28]. Therefore, the above retention curve was defined in part by the substrate particle size distribution that was characterized by a relatively small percentage (≈11%) of particles >8 mm, a large percentage of particles (≈61%) in the range of 8-1 mm and a moderate percentage (≈28%) of particles < 1 mm (Table 2).

Table 2. Particle size distribution of the potted lavender plants' growth substrate Ps$_{96}$:P$_4$ (Ps: pure sphagnum peat and P: perlite in a 96:4 volume ratio).

Particle Size (mm)	Particle Size Distribution (% by wt)
>20	1.24
20–16	1.18
16–10	2.09
10–8	6.70
8–4	18.19
4–2	20.79
2–1	21.85
1–0.5	11.22
0.5–0.25	7.18
0.25–0.106	3.94
0.106–0.053	2.72
<0.053	2.90

3.2. Symptoms Induced by Salinity

Indicator of the effect of salinity are the visible symptoms of the induced damage in plants (such as leaf chlorosis, browning and necrosis) that affects plant ornamental value [29]; an important consideration in floriculture [30]. Throughout the experiment, there was no plant mortality. At the high concentrations of 100 and 200 mM NaCl, all species demonstrated signs of salinity stress expressed initially with chlorosis of the leaves located at the middle and base of the plants followed by gradual necrosis of leaves and stems; at the end of the experiment (day 56), all lavender species irrigated with

200 mM NaCl had leaf necrosis >75% (Table 3). When irrigated with 100 mM NaCl, symptoms of leaf necrosis were more intense in both *Lavandula angustifolia* and *L. stoechas* than in both *L. dentata* var. *dentata* and *L. dentata* var. *candicans*. *L. stoechas* was the first species to demonstrate salinity stress (approximately 28 days since the start of the experiment), followed a week later by *L. angustifolia* and a further week later by both *L. dentata* var. *candicans* and *L. dentata* var. *dentata* (data not shown). On the other hand, all species irrigated with 50 mM NaCl demonstrated moderate signs of salinity stress at completion of the experiment expressed with chlorosis of the leaves (25%–50%) located at the middle and base of the plants; these symptoms reduce the commercial value of the plants. Furthermore, at 25 mM NaCl leaf chlorosis was inconspicuous in all species. Additionally, during the experiment it was noted that only the control of *L. dentata* var. *dentata* demonstrated mild signs of chlorosis suggesting nutrient deficiency; generally, *L. dentata* var. *dentata* developed greater growth (see Section 2.2) in comparison to the other species, therefore, is possible to have had a greater demand for nutrients.

Table 3. Assessment of visual symptoms of potted lavender plants induced by salinity based on a 6-point scale (0–5).

	NaCl Solution				
Species	**0 mM (Control)**	**25 mM**	**50 mM**	**100 mM**	**200 mM**
L. angustifolia	1*	2	3	5	5
L. stoechas	1	2	3	5	5
L. dentata var. *candicans*	1	2	3	4.0	5
L. dentata var. *dentata*	1.5	2	3	4.5	5

where, 0: plant mortality, 1: no leaf injury, 2: mild leaf chlorosis <25%, 3: moderate leaf chlorosis 25%–50% approximately, 4: leaf necrosis 50%-75% approximately, 5: leaf necrosis >75%.

Moderate symptoms of salinity at 50 mM suggest the need for further research to establish the effect of salinity in the range of 25–50 mM NaCl. The use of good quality water between saline water irrigations for strategic leaching of salts could contribute to reducing or delaying the effect of salinity [31]. To further reduce the use of good quality water natural resources for irrigation further research, studying the possibility of strategic leaching of salts with good quality water between saline water irrigations (>50 mM) is also suggested. Overall symptoms were more apparent in *L. dentata* var. *dentata* and *L. stoechas* due to the natural green color foliage of the particular species compared to the natural grey-green foliage of the other two species. Concerning *L. angustifolia* the above symptoms of plants irrigated with 100 and 200 mM NaCl (10.6 dSm^{-1} and 20.7 dSm^{-1}, respectively) were obtained after 56 days of irrigation leading to plant death if the experiment was prolonged further and were similar to the results obtained by Niu and Rodriguez [32] for *L. angustifolia* pot grown plants in peat-perlite based substrate and under greenhouse conditions after 11 weeks of irrigation with NaCl solutions; plants irrigated with 0.8 dS m^{-1} showed good quality with acceptable reduction growth and little leaf injury, whereas plants irrigated with 3.2 dS m^{-1} showed stunted growth and moderate leaf injury (25%–50%), while plants irrigated with either 6.4 dS m^{-1} or 12.0 dS m^{-1} died.

3.3. Plant Growth

With the exception of measurements taken near completion of the experiment (42 and 49 days) in all four species, two-way ANOVA for data concerning the various plant growth variables measured throughout the duration of the experiment showed no significant interactions of the main experimental factors that are between species and irrigated NaCl solutions. In most measured plant variables (plant height, shoot canopy diameter, growth index, Φ_{PSIIo}, inflorescence number) there was a significant effect of the main factors from the beginning of the experiment until day 42. Significant interactions were shown on day 49 and 56 for plant height, day 35 and 56 for Φ_{PSIIo} and at the end of the experiment (day 56) for leaf thickness, inflorescence number and length, peduncle length and root dry weight.

Overall results confirm that salinity stress is initially expressed in the above ground growth parameter of plants [33]. More specifically, plant height was significantly reduced at 200 mM NaCl from day 21 onwards and until day 42 (Table 4). Overall *L. angustifolia* showed the smallest height while *L.* dentata var. *dentata* showed the greatest plant height ($p < 0.05$) between day 7 and 42. Both *L. stoechas* and *L. dentata* var. *candicans* showed similar development in plant height between them that was intermediate of the other two species ($p < 0.05$). Following, near completion of the experiment (days 49 and 56) *L. angustifolia* continued to show the least plant height compared to the other species. On both day 49 and day 56, plant height at 200 mM NaCl was significantly reduced compared to the control in *L. stoechas* and *L. dentata* var. *candicans*. Furthermore, on day 56, plant height at both 100 and 200 mM NaCl was significantly reduced compared to the control in *L. dentata* var. *dentata*. Our results agree with previous research findings that showed salinity induced the decrease in plant height of pot grown *Arbutus unedo* in peat, sand and clay-loam based substrate and irrigated for 16 weeks with 52 mM and 105 mM [34].

Similar results to plant height were obtained for shoot canopy diameter. Shoot canopy diameter was reduced significantly at both 100 and 200 mM NaCl compared to the control from day 49 of the experiment onwards (Table 5). Overall *L. angustifolia* showed the smallest shoot canopy diameter while *L. dentata* var. *dentata* the greatest one throughout the duration of the experiment ($p < 0.05$). *L. dentata* var. *candicans* showed a similar development in shoot canopy diameter with *L. dentata* var. *dentata* until day 35; afterwards, it showed a similar shoot canopy development with *L. stoechas* that was between that of the other two species ($p < 0.05$). Results were also consistent for the plant growth index. Plant growth index was reduced significantly at 200 mM NaCl compared to the control from day 35 of the experiment onwards (Table 6). Overall *L. angustifolia* showed the smallest growth index while *L. dentata* var. *dentata* the greatest one throughout the duration of the experiment ($p < 0.05$). *L. dentata* var. *candicans* showed a similar growth index with *L. dentata* var. *dentata* on day 7. Afterwards, it showed a similar growth index with *L. stoechas* that was between that of the other two species ($p < 0.05$). Similarly, in our results, the canopy and biomass of both *Cistus albidus* and *C. monspeliensis* grown in pots containing peat-pelite based substrate were reduced after 4 months irrigation with 70 and 140 mM NaCl [35]. Also, the canopy and biomass of potted *Asteriscus maritimus* in peat and sand based substrate were reduced after 150 days of irrigation with 70 and 140 mM NaCl [33].

A decrease in shoot dry weight is also an initial effect of reduced growth due to salinity [31]. Shoot dry weight was significantly reduced compared to the control at both 100 and 200 mM NaCl (Table 7). In a similar experiment, the total plant dry weight of *L. multifida* decreased when irrigated with 100 mM and 200 mM NaCl [20,24]. Both *L. dentata* var. *dentata* and *L. dentata* var. *candicans* developed similar shoot dry weights that were significantly greater than the other two species. Furthermore *L. angustifolia* developed the smallest shoot dry weight amongst the studied species ($p < 0.05$). It is possible that there is the presence of some interspecies variation, as the salinity shoot dry weight decrease of the corresponding control at 200 mM was the smallest for *L. angustifolia* (3%), followed in ascending order by *L. dentata* var. *candicans* (18%), *L. dentata* var. *dentata* (27%) and *L. stoechas* (32%). Although *L. angustifolia* (3%) showed the least decrease in shoot dry weight from the corresponding control, the appearance of the plant at the end of the experiment in comparison to the other species exhibited a large amount of leaf and stem necrosis at both 100 mM and 200 mM NaCl (see Section 2.1). The above results agree with the decrease in shoot dry weight due to irrigation with NaCl solutions in other pot grown ornamental species under greenhouse conditions such as *Nerium oleander* [36], *Achillea millefolium*, *Agastache cana*, *Gaillardia aristata* [37], and *Rosmarinus officinalis* [38]. On the other hand, although root dry weight in all species increased at low NaCl concentrations followed by a decrease at 100 and 200 mM NaCl, root dry weights compared to the control were not significantly different. As there were non-significant differences between the control and various NaCl solutions, it is possible that *L. dentata* var. *candicans* developed the greatest root dry weight, while the other species developed similar root dry weights ($p < 0.05$) due to interspecies variation.

Table 4. The effect of irrigation with different NaCl solutions on plant height (cm) of *Lavandula* species (n = 6, $p < 0.05$). Differences between means ± S.E. shown with different letters (Tukey HSD, $p < 0.05$).

Day	7	14	21	28	35	42	49	56
Species								
L. angustifolia	16.467 ± 0.456c	18.047 ± 0.389c	18.583 ± 0.408c	19.063 ± 0.405c	19.257 ± 0.390c	19.267 ± 0.412c	_[†]	-
L. stoechas	25.700 ± 0.456b	28.510 ± 0.389b	29.023 ± 0.408b	29.657 ± 0.405b	30.463 ± 0.390b	31.810 ± 0.412b	-	-
L. dentata var *candicans*	25.583 ± 0.456b	27.410 ± 0.389b	28.207 ± 0.408b	28.900 ± 0.405b	29.280 ± 0.390b	29.623 ± 0.412b	-	-
L. dentata var. *dentata*	29.767 ± 0.456a	33.327 ± 0.389a	34.617 ± 0.408a	36.237 ± 0.405a	37.603 ± 0.390a	38.840 ± 0.412a	-	-
NaCl (mM)								
0 mM (control)	ns	ns	27.779 ± 0.456ab	28.825 ± 0.453a	29.983 ± 0.436a	31.463 ± 0.461a	-	-
25 mM	ns	ns	27.763 ± 0.456ab	28.850 ± 0.453a	29.450 ± 0.436a	30.275 ± 0.461ab	-	-
50 mM	ns	ns	28.525 ± 0.456a	29.392 ± 0.453a	30.271 ± 0.436a	30.829 ± 0.461a	-	-
100 mM	ns	ns	27.467 ± 0.456ab	28.213 ± 0.453ab	28.604 ± 0.436ab	28.904 ± 0.461bc	-	-
200 mM	ns	ns	26.504 ± 0.456b	27.042 ± 0.453b	27.446 ± 0.436b	27.954 ± 0.461c	-	-
Interaction (species x NaCl)								
L. angustifolia x 0 mM (control)	ns	ns	ns	ns	ns	ns	18.817 ± 0.936i	18.817 ± 0.985j
25 mM	ns	ns	ns	ns	ns	ns	19.300 ± 0.936i	19.300 ± 0.985j
50 mM	ns	ns	ns	ns	ns	ns	19.683 ± 0.936i	19.817 ± 0.985j
100 mM	ns	ns	ns	ns	ns	ns	20.083 ± 0.936i	20.083 ± 0.985j
200 mM	ns	ns	ns	ns	ns	ns	18.367 ± 0.936i	18.367 ± 0.985j
L. stoechas x 0 mM (control)	ns	ns	ns	ns	ns	ns	35.950 ± 0.936bcde	36.283 ± 0.985cdef
25 mM	ns	ns	ns	ns	ns	ns	32.483 ± 0.936cdefg	32.950 ± 0.985defgh
50 mM	ns	ns	ns	ns	ns	ns	33.500 ± 0.936cdef	33.750 ± 0.985defg
100 mM	ns	ns	ns	ns	ns	ns	31.400 ± 0.936efgh	31.400 ± 0.985fghi
200 mM	ns	ns	ns	ns	ns	ns	29.567 ± 0.936fgh	29.567 ± 0.985ghi
L. dentata var. *candicans* x 0 mM (control)	ns	ns	ns	ns	ns	ns	32.100 ± 0.936defg	32.517 ± 0.985defgh
25 mM	ns	ns	ns	ns	ns	ns	30.600 ± 0.936fgh	30.600 ± 0.985ghi
50 mM	ns	ns	ns	ns	ns	ns	31.633 ± 0.936efgh	31.633 ± 0.985efgh
100 mM	ns	ns	ns	ns	ns	ns	27.950 ± 0.936gh	27.950 ± 0.985hi
200 mM	ns	ns	ns	ns	ns	ns	27.183 ± 0.936h	26.467 ± 0.985i
L. dentata var. *dentata* x 0 mM (control)	ns	ns	ns	ns	ns	ns	41.600 ± 0.936a	42.183 ± 0.985a
25 mM	ns	ns	ns	ns	ns	ns	40.917 ± 0.936a	41.500 ± 0.985ab
50 mM	ns	ns	ns	ns	ns	ns	40.333 ± 0.936ab	40.333 ± 0.985abc
100 mM	ns	ns	ns	ns	ns	ns	36.833 ± 0.936abcd	36.933 ± 0.985bcd
200 mM	ns	ns	ns	ns	ns	ns	36.983 ± 0.936abc	36.517 ± 0.985bcde
$F_{species}$/sig.	151.694/0.000*	270.206/0.000*	266.763/0.000*	305.079/0.000*	375.767/0.000*	386.198/0.000*	397.793/0.000	364.676/0.000
F_{NaCl}/sig.	2.158/0.079	1.641/0.170	2.568/0.043*	3.926/0.005*	6.909/0.000*	9.685/0.000*	12.834/0.000	14.908/0.000
$F_{interaction}$/sig.	0.981/0.472	1.135/0.341	0.740/0.709	0.695/0.753	1.272/0.247	1.374/0.191	1.856/0.049*	1.961/0.036*

ns: non-significant; * denotes significant differences between means at $p < 0.05$, shown with different letters within columns. [†]When interactions are significant, factors are not considered and mean values are not shown.

Table 5. The effect of irrigation with different NaCl solutions on shoot canopy diameter (cm) of *Lavandula* species (n = 6, $p < 0.05$). Differences between means ± S.E. shown with different letters (Tukey HSD, $p < 0.05$).

Day	7	14	21	28	35	42	49	56
Species								
L. angustifolia	15.940 ± 0.413c	18.747 ± 0.441c	19.473 ± 0.417c	19.903 ± 0.418c	20.037 ± 0.412c	20.407 ± 0.415c	20.510 ± 0.418c	20.533 ± 0.419c
L. stoechas	21.890 ± 0.413b	26.647 ± 0.441b	27.600 ± 0.417b	28.330 ± 0.418b	29.173 ± 0.412b	29.950 ± 0.415b	30.417 ± 0.418b	30.417 ± 0.419b
L. dentata var. *candicans*	27.233 ± 0.413a	31.457 ± 0.441a	32.873 ± 0.417a	33.840 ± 0.418a	34.097 ± 0.412a	34.427 ± 0.415b	34.473 ± 0.418b	34.473 ± 0.419b
L. dentata var. *dentata*	26.350 ± 0.413a	31.750 ± 0.441a	32.603 ± 0.417a	34.477 ± 0.418a	35.203 ± 0.412a	37.160 ± 0.415a	37.797 ± 0.418a	37.863 ± 0.419a
NaCl (mM)								
0 mM (control)	ns	ns	ns	ns	ns	ns	32.021 ± 0.467a	32.117 ± 0.469a
25 mM	ns	ns	ns	ns	ns	ns	31.063 ± 0.467ab	31.079 ± 0.469ab
50 mM	ns	ns	ns	ns	ns	ns	31.125 ± 0.467ab	31.125 ± 0.469ab
100 mM	ns	ns	ns	ns	ns	ns	30.163 ± 0.467b	30.163 ± 0.469b
200 mM	ns	ns	ns	ns	ns	ns	29.625 ± 0.467b	29.625 ± 0.469b
$F_{species}$/sig.	156.673/0.000*	189.365/0.000*	226.035/0.000*	260.722/0.000*	280.935/0.000*	313.565/0.000*	321.313/0.000*	320.620/0.000*
F_{NaCl}/sig.	1.289/0.280	0.536/0.710	0.395/0.811	1.518/0.203	1.315/0.270	1.933/0.111	3.950/0.005*	4.215/0.003*
$F_{interaction}$/sig.	1.302/0.229	0.243/0.995	0.382/0.967	1.305/0.228	0.556/0.872	0.528/0.892	0.635/0.808	0.656/0.789

ns: non-significant; * denotes significant differences between means at $p < 0.05$, shown with different letters within columns.

Table 6. The effect of irrigation with different NaCl solutions on plant growth index of *Lavandula* species (n = 6, $p < 0.05$). Differences between means ± S.E. shown with different letters (Tukey HSD, $p < 0.05$).

Day	7	14	21	28	35	42	49	56
Species								
L. angustifolia	16.100 ± 0.335c	18.493 ± 0.323c	19.157 ± 0.318c	19.617 ± 0.327c	19.767 ± 0.309c	20.010 ± 0.318c	20.073 ± 0.329c	20.097 ± 0.333c
L. stoechas	23.153 ± 0.335b	27.250 ± 0.323b	28.067 ± 0.318b	28.760 ± 0.327b	29.593 ± 0.309b	30.553 ± 0.318b	31.110 ± 0.329b	31.187 ± 0.333b
L. dentata var. *candicans*	26.687 ± 0.335a	30.087 ± 0.323b	31.297 ± 0.318b	32.173 ± 0.327b	32.480 ± 0.309b	32.807 ± 0.318b	32.927 ± 0.329b	32.910 ± 0.333b
L. dentata var. *dentata*	27.483 ± 0.335a	32.263 ± 0.323a	33.253 ± 0.318a	34.720 ± 0.327a	35.993 ± 0.309a	37.710 ± 0.318a	38.310 ± 0.329a	38.407 ± 0.333a
NaCl (mM)								
0 mM (control)	ns	ns	ns	ns	30.296 ± 0.345a	31.300 ± 0.356a	32.038 ± 0.367a	32.213 ± 0.372a
25 mM	ns	ns	ns	ns	29.617 ± 0.345ab	30.492 ± 0.356ab	30.967 ± 0.367ab	31.071 ± 0.372ab
50 mM	ns	ns	ns	ns	29.821 ± 0.345ab	30.825 ± 0.356ab	31.171 ± 0.367ab	31.204 ± 0.372ab
100 mM	ns	ns	ns	ns	28.996 ± 0.345ab	29.717 ± 0.356bc	29.775 ± 0.367bc	29.783 ± 0.372bc
200 mM	ns	ns	ns	ns	28.563 ± 0.345b	29.017 ± 0.356c	29.075 ± 0.367c	28.979 ± 0.372c
$F_{species}$/sig.	239.689/0.000*	351.413/0.000*	384.017/0.000*	406.690/0.000*	224.500/0.000*	550.696/0.000*	542.787/0.000*	532.256/0.000*
F_{NaCl}/sig.	0.464/0.762	0.138/0.968	0.948/0.440	2.267/0.067	3.933/0.049*	6.514/0.000*	10.242/0.000*	11.675/0.000*
$F_{interaction}$/sig.	0.862/0.587	0.431/0.948	0.619/0.822	0.962/0.490	0.922/0.528	0.860/0.590	1.139/0.338	1.256/0.257

ns: non-significant; * denotes significant differences between means at $p < 0.05$, shown with different letters within columns.

Table 7. The effect of irrigation with different NaCl solutions on shoot and root dry weights (g) of *Lavandula* species (n = 6, $p < 0.05$). Differences between means ± S.E. shown with different letters (Tukey HSD, $p < 0.05$).

		Shoot	Root
Species			
L. angustifolia		20.638 ± 0.950c	-[†]
L. stoechas		34.428 ± 0.950b	-
L. dentata var. *candicans*		47.178 ± 0.950a	-
L. dentata var. *dentata*		43.854 ± 0.650a	-
NaCl (mM)			
0 mM (control)		41.270 ± 1.062a	-
25 mM		36.907 ± 1.062 b	-
50 mM		37.534 ± 1.062ab	-
100 mM		34.974 ± 1.062bc	-
200 mM		31.937 ± 1.062c	-
Interaction (species x NaCl)			
L. angustifolia x	0 mM (control)	ns	5.917 ± 0.501cd
	25 mM	ns	6.282 ± 0.501c
	50 mM	ns	4.865 ± 0.501cd
	100 mM	ns	4.985 ± 0.501cd
	200 mM	ns	5.123 ± 0.501cd
L. stoechas x	control	ns	5.593 ± 0.501cd
	25 mM	ns	6.417 ± 0.501c
	50 mM	ns	4.752 ± 0.501cd
	100 mM	ns	5.575 ± 0.501cd
	200 mM	ns	3.488 ± 0.501d
L. dentata var. *candicans* x	0 mM (control)	ns	13.887 ± 0.501ab
	25 mM	ns	14.847 ± 0.501ab
	50 mM	ns	15.180 ± 0.501a
	100 mM	ns	15.422 ± 0.501a
	200 mM	ns	12.277 ± 0.501b
L. dentata var. *dentata* x	0 mM (control)	ns	5.315 ± 0.501cd
	25 mM	ns	4.800 ± 0.501cd
	50 mM	ns	5.533 ± 0.501cd
	100 mM	ns	4.667 ± 0.501cd
	200 mM	ns	4.703 ± 0.501cd
$F_{species}$/**sig.**		156.577/0.000*	415.648/0.000
F_{NaCl}/**sig.**		10.444/0.000*	6.470/0.000
$F_{interaction}$/**sig.**		1.312/0.223	2.396/0.009*

ns: non-significant; * denotes significant differences between means at $p < 0.05$, shown with different letters within columns. [†]When interactions are significant, factors are not considered, and mean values are not shown.

Leaf thickening is a common response to salinity [31,39], whereby salt content in the leaves is diluted by increased succulence [40]. Leaf thickness increased significantly compared to the control only in *L. angustifolia* at 50 mL NaCl (Figure 3); however, non-significant leaf thickness increase compared to the control was generally observed at concentrations >50 mM NaCl in the remaining lavender species. It is possible over the 56 days, NaCl levels, especially in the higher concentrations (100 and 200 mM NaCl), exceeded the threshold of long-term acclimation mechanisms related to leaf thickness. Further research with smaller NaCl levels is necessary to study the effect of salinity on leaf thickness.

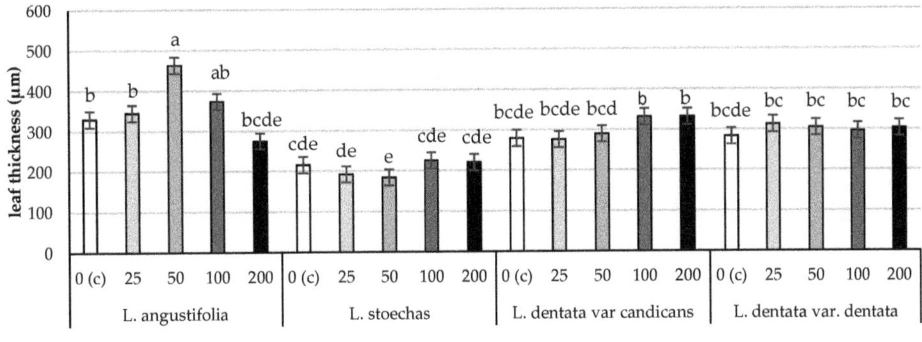

Figure 3. The combined effect of the interaction between lavender species and NaCl solution irrigation treatments (0, 25, 50, 100, 200 mM) on leaf thickness of plants (n = 6, $p < 0.05$). Differences between means ± S.E. shown with different letters (Tukey HSD, $p < 0.05$). Note c: control.

A secondary effect of reduced growth induced by salinity includes the potential reduction in photosynthesis of plant leaves [16]. Concerning Φ_{PSIIo}, *L. dentata* var. *dentata* showed the least Φ_{PSIIo} compared to the other species ($p < 0.05$) on day 14 (Table 8), suggesting the presence of interspecies variation in plant nutrient demand, as there were non-significant differences between the control and the different NaCl solutions applied and the fact that *L. dentata* var. *dentata* Φ_{PSIIo} values in the control (0.77) were relatively less than optimum (circa 0.83) [41]. With regards to the latter, *L. dentata* var. *dentata* also showed mild signs of chlorosis in the control, suggesting a greater demand for nutrients in comparison to the other species (see Section 2.1). Following this, Φ_{PSIIo} significantly decreased in plants irrigated with 200 mM NaCl compared to the control on day 35 only in *L. angustifolia*, and on day 56 in both *L. angustifolia* and *L. dentata* var. *candicans*. The Φ_{PSIIo} values for all species irrigated with 200 NaCl was lower than the optimum Φ_{PSIIo} value for most plant species (circa 0.83), indicating the exposure of plants to stress, in this study salt stress [41]. Although no significant differences were shown, the remaining Φ_{PSIIo} values of plants irrigated with either 100 mM or 200 mM were also less than the optimum Φ_{PSIIo} in all species. The decrease in Φ_{PSIIo} suggests the presence of salt stress that led to chlorosis and premature senescence of mature leaves [42]. These results agree with the symptoms or leaf chlorosis and necrosis observed on the plants irrigated with 100 and 200 mM NaCl and also with Munns' [43] findings that high potassium concentrations cause premature senescence, chlorosis, and necrosis in leaves due to the disrupt of plant protein synthesis.

Plants subjected to salt stress could reduce the inflorescence number, influence flowering time (speed or delay) and duration, reduce inflorescence and reduce peduncle length. Throughout the duration of the experiment, only two lavender species flowered (*L. stoechas* and *L. dentata* var. *dentata*). The experiment took place in late winter–early spring (Feb-March) and as the control of both non-flowered lavenders (*L. angustifolia* and *Lavandula dentata* var. *candicans*) did not produce inflorescences, it is possible that the season in which the experiment took place was too early for these species to flower. It is reported that *L. dentata* var. *dentata* flower all year round in mild-winter areas such as Athens [44] and *L. stoechas* flowers in early spring [45], whereas *L. dentata* var. *candicans* from early spring to late fall and *L. angustifolia* flowers from early to midsummer [44]. Anthesis in both *L. stoechas* and *L. dentata* var. *dentata* started earlier at 100 and 200 mM NaCl compared to the control (Figure 4). At the end of the experiment (day 56), the inflorescence number for the two lavender species that flowered was reduced significantly at 100 and 200 mM in comparison to the control only for *L. stoechas*, suggesting a tendency for earlier inflorescence death under salt stress. Similarly, the inflorescence length was reduced significantly at 100 and 200 mM only for *L. stoechas*. On the other hand, the inflorescence peduncle was reduced significantly compared to the control in all NaCl concentrations for both flowered *L. stoechas* and *L. dentata* var. *dentata*. The above results agree with the findings of several authors that have studied the effect of salinity on various ornamental plant species. García-Caparrós and Lao [46] state that salinity could bring forward and shorten the duration of anthesis. Salinity has been reported to decrease the inflorescence number in gerbera [47,48], different cultivars of *Rosa* x *hybrida* L. [48,49], *Matricaria chamomilla* [50], inflorescence length in *Eustoma grandiflorum* [18] as well as the peduncle length / stem length (measured from the basis of the plant to the first flower) in *Dianthus caryophyllus*, *Gerbera jamesonii* L [47], *Eustoma grandiflorum* [18], *Rosa hybrida* 'Kardinal' [51], and *Matricaria chamomilla* [50]. Results also showed that *Lavandula stoechas* developed significantly more inflorescences than *L. dentata* var. *dentata* on both day 42 and 49, which is possibly due to interspecies differences, as there were no differences between the control and applied NaCl solutions.

Table 8. The effect of irrigation with different NaCl solutions on leaf Φ_{PSIIo} of *Lavandula* species (n = 6, $p < 0.05$). Differences between means ± S.E. shown with different letters (Tukey HSD, $p < 0.05$).

	Day	**14**	**35**	**56**
Species				
L. angustifolia		0.821 ± 0.007a	-[†]	-
L. stoechas		0.807 ± 0.007a	-	-
L. dentata var. *candicans*		0.810 ± 0.007a	-	-
L. dentata var. *dentata*		0.755 ± 0.007b	-	-
NaCl (mM)				
0 mM (control)		ns	-	-
25 mM		ns	-	-
50 mM		ns	-	-
100 mM		ns	-	-
200 mM		ns	-	-
Interaction (species x NaCl)				
L. angustifolia x	0 mM (control)	ns	0.788 ± 0.056a	0.821 ± 0.075a
	25 mM	ns	0.816 ± 0.056a	0.833 ± 0.075a
	50 mM	ns	0.789 ± 0.056a	0.794 ± 0.075a
	100 mM	ns	0.674 ± 0.056a	0.500 ± 0.075abc
	200 mM	ns	0.160 ± 0.056b	0.215 ± 0.075bc
L. stoechas x	0 mM (control)	ns	0.815 ± 0.056a	0.832 ± 0.075a
	25 mM	ns	0.790 ± 0.056a	0.798 ± 0.075a
	50 mM	ns	0.797 ± 0.056a	0.775 ± 0.075a
	100 mM	ns	0.723 ± 0.056a	0.470 ± 0.075abc
	200 mM	ns	0.566 ± 0.056a	0.493 ± 0.075abc
L. dentata var. *candicans* x	0 mM (control)	ns	0.821 ± 0.056a	0.807 ± 0.075a
	25 mM	ns	0.820 ± 0.056a	0.821 ± 0.075a
	50 mM	ns	0.796 ± 0.056a	0.828 ± 0.075a
	100 mM	ns	0.748 ± 0.056a	0.631 ± 0.075a
	200 mM	ns	0.728 ± 0.056a	0.130 ± 0.075c
L. dentata var. *dentata* x	0 mM (control)	ns	0.770 ± 0.056a	0.788 ± 0.075a
	25 mM	ns	0.743 ± 0.056a	0.731 ± 0.075a
	50 mM	ns	0.717 ± 0.056a	0.767 ± 0.075a
	100 mM	ns	0.632 ± 0.056a	0.585 ± 0.075ab
	200 mM	ns	0.660 ± 0.056a	0.523 ± 0.075ab
$F_{species}$/**sig.**		19.181/0.000*	5.358/0.000	0.450/0.718
F_{NaCl}/**sig.**		0.352/0.842	16.579/0.000	30.580/0.000
$F_{interaction}$/**sig.**		1.061/0.401	4.322/0.000*	1.959/0.036*

ns: non-significant; * denotes significant differences between means at $p < 0.05$, shown with different letters within columns. [†] When interactions are significant, factors are not considered, and mean values are not shown.

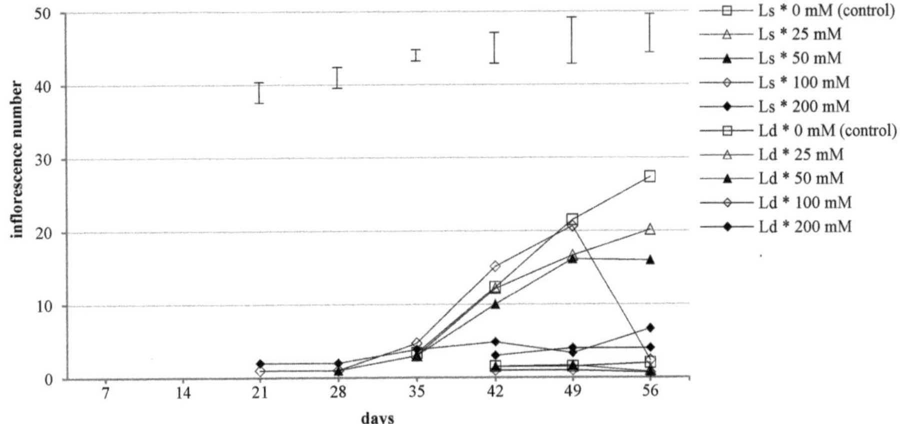

Figure 4. The combined effect of the interaction between lavender species and NaCl solution irrigation treatments (0, 25, 50, 100, 200 mM) on plant inflorescence number (n = 6, $p < 0.05$). Note: Ls: *L. stoechas*, Ld: *L. dentata* var. *dentata*. Bars represent S.E.

Considering all of the above, the overall performance of the studied lavender species under the effect of salinity was satisfactory at levels <25 mM NaCl in all species. Further research is necessary to establish the effect of salinity between 25–50 mM NaCl or possibly >50 mM in combination with the interchanged use of irrigation with good quality water with the aim to contribute towards the conservation of good quality water natural resources. The adverse effects induced by high levels of NaCl (>100 mM) amongst lavender species in ascending order was *Lavandula dentata* var. *dentata*, *L. dentata* var. *candicans*, *L. stoechas* and *L. angustifolia*. Therefore, in areas with saline irrigation water, the use of the better performed in the current study under saline irrigation conditions *Lavandula dentata* var. *dentata* and *L. dentata* var. *candicans* is proposed.

4. Conclusions

The study of the effect of saline irrigation in floriculture is important to consider for producing nursery crops without signs of salinity injury. Amongst other factors, the level of salinity stress induced on plants is dependent on plant species and varieties. The effect of salinity through irrigation on the growth of four lavender species was determined. The applied irrigation method allowed plants to receive the adequate amount of water for plant growth (easily available water), ensuring the effect of irrigation was induced by water quality i.e., applied NaCl solutions.

The effect of salinity was initially expressed in most of the above ground growth variables of the plants studied. Generally, growth was satisfactory in all species irrigated with <25 mM NaCl. Symptoms of salinity injury were moderate at 50 mM NaCl, affecting the commercial value of the ornamental species. At high NaCl levels (100 and 200 mM), plants showed severe symptoms of salt stress that included leaf and stem necrosis. Only two lavender species flowered, possibly due to season variation between species. Anthesis time was quicker and anthesis duration was reduced for plants irrigated with high saline concentrations.

Overall results of the effect of salinity were consistent, allowing to rank species in descending order of plant development as follows: *Lavandula dentata* var. *dentata*, *L. dentata* var. *candicans*, *L. stoechas* and *L. angustifolia*. Throughout the duration of the experiment, both *Lavandula dentata* var. *dentata* and *L. dentata* var. *candicans* showed better growth, and hence are suggested for areas with poor water quality using saline water for irrigation.

Author Contributions: Conceptualization, A.T.P.; methodology, A.T.P., P.A.L. and G.L.; formal analysis, A.T.P. A.K.K. and K.B.; investigation, A.K.K. and A.T.P.; writing—original draft preparation, A.T.P.; writing—review and editing, A.T.P., P.A.L., G.L. and K.B.; supervision, A.T.P., P.A.L. and G.L.; project administration, A.T.P. A.K.K. and K.B. All authors have read and agreed to the published version of the manuscript.

Funding: This research received no external funding.

Acknowledgments: We would like to thank Kalantzis Plants for the free supply of lavender plants.

Conflicts of Interest: The authors declare no conflict of interest.

References

1. Daniele, L.; Vallejos, A.; Sola, F.; Corbella, M.; Pullido-Bosch, A. Hydrogeochemical processes in the vicinity of a desalination plant (Cabo de Gata, SE Spain). *Desalination* **2011**, *277*, 338–347. [CrossRef]

2. Alfarrah, N.; Walraevens, K. Groundwater Overexploitation and Seawater Intrusion in Coastal Areas of Arid and Semi-Arid Regions. *Water* **2018**, *10*, 143. [CrossRef]

3. Liakou, E. Θεσμικό πλαίσιο για την επαναχρησιμοποίηση επεξεργασμένων υγρών αστικών αποβλήτων (*Legislative Framework For the Reuse of Treated Urban Waste Water*); Ημερίδα, Επαναχρησιμοποίηση αστικών λυμάτων για άρδευση, Θεσμικό πλαίσιο, εφαρμογέσ και προοπτικέσ για την Ελλάδα. Δ.Ε.Υ.Α.Χ. και Πολυτεχνείο Κρήτησ: Χανιά, Ελλάδα, 2013.

4. Cai, X.; Sun, Y.; Starman, T.; Hall, C. Response of 18 Earth-Kind Rose Cultivars to Salt Stress. *HortScience* **2014**, *49*, 544–549. [CrossRef]

5. Pessarakli, M.; Szabolcs, I. Soil salinity and sodicity as particular plant/crop stress factors. In *Handbook of Plant and Crop Stress*, 3rd ed.; Pessarakli, M., Ed.; CRC Press Taylor and Francis Group: Boca Raton, FL, USA, 2011; pp. 3–21.

6. Karabourniotis, G.; Liakopoulos, G.; Nikolopoulos, D. Φυσιολογία Καταπονήσεων των Φυτών: Οι λειτουργίες των φυτών κάτω από αντίξοες συνθήκες του περιβάλλοντος (*Plant Stress Physiology: Plant Functioning under Environmental Stress*); Εκδόσεισ Έμβρυο: Egaleo, Greece, 2012.

7. Lambers, H. Dryland salinity: A key environmental issue in southern Australia. *Plant Soil* **2003**, *257*. [CrossRef]

8. Batool, N.; Shahzad, A.; Ilyas, N. Plants and Salt stress. *Int. J. Agric. Crop. Sci.* **2014**, *7*, 1439–1446.

9. Parida, A.K.; Das, A.B. Salt tolerance and salinity effects on plants: A review. *Ecotox. Environ. Saf.* **2005**, *60*, 324–349. [CrossRef] [PubMed]

10. James, R.A.; Rivelli, A.R.; Munns, R.; von Caemmerer, S. Factors affecting CO_2 assimilation, leaf injury and growth in salt-stressed durum wheat. *Fun. Plant Biol.* **2002**, *29*, 1393–1403. [CrossRef]

11. Munns, R. Genes and salt tolerance: Bringing them together. *New Phytol.* **2005**, *167*, 645–663. [CrossRef]

12. Bernstein, L. Effects of salinity and sodicity on plant growth. *Ann. Rev. Phytopathol.* **1975**, *13*, 295–312. [CrossRef]

13. Nikaya, A.; Masui, M.; Ishida, A. Salt tolerance of muskmelons in sand nutrient solution cultures. *J. Jpn. Soc. Hort. Sci.* **1983**, *49*, 93–101. [CrossRef]

14. Carillo, P.; Annunziata, M.G.; Pontecorvo, G.; Fuggi, A.; Woodrow, P. Salinity Stress and Salt Tolerance. In *Abiotic Stress in Plants–Mechanisms and Adaptations*; Shanker, A., Ed.; Tech: Rijeka, Croatia, 2011; pp. 21–38.

15. Demiral, M.A. Effect of salt stress on concentration of nitrogen and phosphorus in root and leaf of strawberry plant. *Eurasian J. Soil Sci.* **2017**, *6*, 357–364. [CrossRef]

16. Greenway, H.; Munns, R. Mechanisms of salt tolerance in nonhalophytes. *Ann. Rev. Plant Physiol.* **1980**, *31*, 149–190. [CrossRef]

17. Grieve, C.M. Review: Irrigation of floricultural and nursery crops with saline wastewaters. *Isr. J. Plant Sci.* **2011**, *59*, 187–196. [CrossRef]

18. Shillo, R.; Ding, M.; Pasternak, D.; Zaccai, M. Cultivation of cut flower and bulb species with saline water. *Sci. Hortic.* **2002**, *92*, 41–54. [CrossRef]

19. Fornes, F.; Belda, R.M.; Carrión, C.; Noguera, V.; García-Agustín, P.; Abad, M. Pre-conditioning ornamental plants to drought by means of saline water irrigation as related to salinity tolerance. *Sci. Hortic.* **2007**, *113*, 52–59. [CrossRef]

20. Plaza, B.M.; Jiménez-Becker, S.; García-Caparrós, P.; del M. Verdejo, M. Influence of Salinity on Vegetative Growth of Six Native Mediterranean Species. *Acta Hortic.* **2015**, *1099*, 755–760. [CrossRef]

21. The Plant List (2013). Version 1.1. Available online: http://www.theplantlist.org/1.1/browse/A/Lamiaceae/ Lavandula/ (accessed on 6 January 2020).

22. Schaminée, J.H.J.; Chytrý, M.; Hennekens, S.M.; Janssen, J.A.M.; Jiménez-Alfaro, B.; Knollová, I.; Marceno, C.; Mucina, L.; Rodwell, J.S.; Tichý, L. *Review of Grassland Habitats and Development of Distribution Maps of Heathland, Scrub and Tundra Habitats of EUNIS Habitats Classification*; Report EEA/NSV/15/005; Stichting Dienst Landbouwkundig Onderzoek–Alterra (ALT): Wageningen, The Netherlands, 2016.

23. Polunin, O. *Flowers of Greece and the Balkans, a Field Guide*; Oxford University Press: New York, NY, USA, 1997.

24. García-Caparrós, P.; Llanderal, A.; Pestana, M.; Correia, P.J.; Lao, M.T. Lavandula multifida response to salinity: Growth, nutrient uptake, and physiological changes. *J. Plant Nutr. Soil Sci.* **2016**, 1–9. [CrossRef]

25. Haines, W.B. Studies in the physical properties of soils. V. The hysteresis effect in capillary properties and the modes of moisture distribution associated therewith. *J. Agric. Sci.* **1930**, *20*, 97–116. [CrossRef]

26. De Boodt, M.; Verdonck, O. The physical properties of the substrates in horticulture. *Acta Hortic.* **1972**, *26*, 37–44. [CrossRef]

27. Raviv, M.; Lieth, J.H. *Soilless Culture Theory and Practice*; Elsevier BV: London, UK, 2008; p. 587.

28. Naaz, R.; Bussières, P. Particle Sizes Related to Physical Properties of Peat-Based Substrates. *Acta Hortic.* **2011**, *893*, 971–978. [CrossRef]

29. Caser, M.; Scariot, V.; Gaino, W.; Larcher, F.; Devecchi, M. The Effects of Sodium Chloride on the Aesthetic Value of Buxus spp. *Eur. J. Hort. Sci.* **2013**, *78*, 153–159.

30. Cassaniti, C.; Leonardi, C.; Flowers, T.J. The effect of sodium chloride on ornamental shrubs. *Sci. Hortic.* **2009**, *122*, 586–593. [CrossRef]

31. Cassaniti, C.; Romano, D.; Flowers, T.J. The Response of Ornamental Plants to Saline Irrigation Water. In *Irrigation-Water Management, Pollution and Alternative Strategies*; Garcia-Garizabal, I., Ed.; Tech: London, UK, 2012; pp. 131–158.

32. Niu, G.; Rodriguez, D.S. Relative salt tolerance of selected herbaceous perennials and groundcovers. *Sci. Hortic.* **2006**, *110*, 352–358. [CrossRef]

33. Rodríguez, P.; Torrecillas, A.; Morales, M.A.; Ortuño, M.F.; Sánchez-Blanco, M.J. Effects of NaCl salinity and water stress on growth and leaf water relations of *Asteriscus maritimus* plants. *Environ. Exp. Bot.* **2005**, *53*, 113–123. [CrossRef]

34. Navarro, A.; Bañon, S.; Olmos, E.; Sánchez-Blanco, M.J. Effect of sodium chloride on water potential components, hydraulic conductivity, gas exchange and leaf ultrastructure of *Arbutus unedo* plants. *Plant Sci.* **2007**, *172*, 473–480. [CrossRef]

35. Torrecillas, A.; Rodríguez, P.; Sánchez-Blanco, M.J. Comparison of growth, leaf water relations and gas exchange of *Cistus albidus* and *C. monspeliensis* plants irrigated with water of different NaCl salinity levels. *Sci. Hortic.* **2003**, *97*, 353–368. [CrossRef]

36. Bañón, S.; Fernández, J.A.; Ochoa, J.; Sánchez-Blanco, M.J. Paclobutrazol as an aid to reduce some effects of salt stress in oleander seedlings. *Eur. J. Hort. Sci.* **2005**, *70*, 43–49.

37. Niu, G.; Rodriguez, D.S. Relative salt tolerance of five herbaceous perennials. *Hortscience* **2006**, *41*, 1493–1497. [CrossRef]

38. Alarcón, J.J.; Morales, M.A.; Ferrández, T.; Sánchez-Blanco, M.J. Effects of water and salt stress on growth, water relations and gas exchange in *Rosmarinus officinalis*. *J. Hortic. Sci. Biotech.* **2006**, *81*, 845–853. [CrossRef]

39. Imbrahim, K.M.; Collins, J.C.; Collin, H.A. Effects of salinity on growth and ionic composition of *Coleus blumei* and *Salvia splendens*. *J. Hortic. Sci.* **2015**, *66*, 215–222. [CrossRef]

40. Luttge, U.; Smith, J.A.C. Structural, biophysical and biochemical aspects of the role of leaves in plant adaption to salinity and water stress. In *Salinity Tolerance in Plant. Strategies for Crop Improvement*; Staples, R.C., Toenniessen, G.H., Eds.; Wiley International: New York, NY, USA, 1984; pp. 125–150.

41. Maxwell, K.; Johnson, G.N. Chlorophyll fluorescence, a practical guide. *J. Exp. Bot.* **2000**, *51*, 659–668. [CrossRef] [PubMed]

42. Sultana, N.; Ikeda, T.; Itoh, R. Effect of NaCl salinity photosynthesis and dry matter accumulation in developing rice grains. *Environ. Exp. Bot.* **1999**, *42*, 211–220. [CrossRef]

43. Munns, R. Comparative physiology of salt and water stress. *Plant Cell Environ.* **2002**, *25*, 239–250. [CrossRef] [PubMed]

44. Brenzel, K.N. *Sunset Western Garden Book*; Sunset Publishing Corporation: Oakland, CA, USA, 2007; p. 768.

45. Sfikas, G. *Αγριολούλουδα της Κρήτης* (*Wildflowers of Crete*); Ευςταθιάδης και Υιοί A.E.: Αθήνα, Ελλάδα, 1987; p. 310.

46. García-Caparrós, P.; Lao, M.T. The effects of salt stress on ornamental plants and integrative cultivation practices. *Sci. Hortic.* **2018**, *240*, 430–439. [CrossRef]

47. Baas, R.; Nijssen, H.M.C.; van den Berg, T.J.M.; Warmenhoven, M.G. Yield and quality of carnation (*Dianthus caryophyllus* L.) and gerbera (*Gerbera jamesonii* L.) in a closed nutrient system as affected by sodium chloride. *Sci. Hortic.* **1995**, *61*, 273–284. [CrossRef]

48. Sonneveld, C.; Baas, R.; Nijssen, H.M.C.; de Hoog, J. Salt Tolerance of Flower Crops Grown in Soilless Culture. *J. Plant Nutr.* **1999**, *22*, 1033–1048. [CrossRef]

49. Cai, X.; Niu, G.; Starman, T.; Hall, C. Response of six garden roses (*Rosa* x *hybrida* L.) to salt stress. *Sci. Hortic.* **2014**, *168*, 27–32. [CrossRef]

50. Rajzmoo, K.; Heydarizadeh, P.; Sabzalian, M.R. Effect of Salinity and Drought Stresses on Growth Parameters and Essential Oil Content of *Matricaria chamomila*. *Int. J. Agric. Biol.* **2008**, *10*, 451–454.

51. Wahome, P.K.; Jesch, H.H.; Grittner, I. Effect of NaCl on the vegetative growth and flower quality of roses. *J. Appl. Bot.* **2000**, *74*, 38–41.

 © 2020 by the authors. Licensee MDPI, Basel, Switzerland. This article is an open access article distributed under the terms and conditions of the Creative Commons Attribution (CC BY) license (http://creativecommons.org/licenses/by/4.0/).

Article

Dynamics and Distribution of Soil Salinity under Long-Term Mulched Drip Irrigation in an Arid Area of Northwestern China

Zilong Guan [1,2,3], Zhifeng Jia [1,2,3,*], Zhiqiang Zhao [1,3] and Qiying You [1,2,3]

[1] School of Environmental Science and Engineering, Chang'an University, Xi'an 710054, China; hydgeo_guan@163.com (Z.G.); zzq8055@sina.com (Z.Z.); qiyingyou@163.com (Q.Y.)
[2] Institute of Water and Development, Chang'an University, Xi'an 710054, China
[3] Key Laboratory of Subsurface Hydrology and Ecological Effects in Arid Region, Ministry of Education, Chang'an University, Xi'an 710054, China
*   Correspondence: 409538088@chd.edu.cn; Tel.: +86-29-8233 9323

Received: 8 May 2019; Accepted: 9 June 2019; Published: 12 June 2019

Abstract: Mulched drip irrigation has been widely used in agricultural planting in arid and semi-arid regions. The dynamics and distribution of soil salinity under mulched drip irrigation greatly affect crop growth and yield. However, there are still different views on the distribution and dynamics of soil salinity under long-term mulched drip irrigation due to complex factors (climate, groundwater, irrigation, and soil). Therefore, the soil salinity of newly reclaimed salt wasteland was monitored for 9 years (2008–2016), and the effects of soil water on soil salinity distribution under mulched drip irrigation have also been explored. The results indicated that the soil salinity decreased sharply in 3–4 years of implementation of mulched drip irrigation, and then began to fluctuate to different degrees and showed slight re-accumulation. During the growth period, soil salinity was relatively high at pre-sowing, and after a period of decline soil salinity tends to increase in the late harvest period. The vertical distribution of soil texture had a significant effect on the distribution of soil salinity. Salt accumulated near the soil layer transiting from coarse soil to fine soil. After a single irrigation, the soil water content in the 30–70 cm layer under the cotton plant undergoes a 'high–low–high' change pattern, and the soil salt firstly moved to the deep layer (below 70 cm), and then showed upward migration tendency with the weakening of irrigation water infiltration. The results may contribute to the scientific extension of mulched drip irrigation and the farmland management under long-term mulched drip irrigation.

Keywords: arid zone; soil salinity; long-term mulched drip irrigation; soil texture; cotton

---

1. Introduction

Soil salinization is a major obstacle to the sustainable development of irrigated agriculture. Ten percent of arable lands worldwide are affected by salinization, and 4 million km^2 of arable lands lose planting function due to salinization [1,2]. Secondary salinization has a more serious impact on agricultural development in arid and semi-arid areas, with salinization affecting 9–25% of cultivated land in Tunisia, the United States, India and South Africa. Xinjiang, Northwest China, suffers from severe soil salinity. Saline-alkali land in Xinjiang is accounting for nearly one third of the total saline-alkali land in China, and most of the saline-alkali lands are within the oasis [3]. Cotton, one of the main crops in Xinjiang, belongs to a salt-tolerant crop. The cotton tolerance to irrigation water salinity (electrical conductivity) and soil salinity (electrical conductivity) are 5.1 dS m^{-1} and 7.7 dS m^{-1} under 100% yield potential, while 18 dS m^{-1} and 27 dS m^{-1} under 0% yield potential [4]. The large area of severe saline-alkali land seriously restricts effective use of the land and the development of agriculture [5]. However, soil salt redistribution is accompanied by irrigation in arid areas, and different irrigation methods will lead to different water and salt transport and distribution [6]. On the one hand, irrigation water can leach the salt in the soil and dilute the concentration of the soil solution. On the other hand, unreasonable irrigation or poor drainage will result in a rapid rise of the groundwater table, which could lead to soil salinization. For irrigated areas in arid regions, secondary soil salinization is a high probability event under the combined action of groundwater level, groundwater mineralization, soil texture [7,8]. Therefore, effective tillage and irrigation methods are essential to mitigate the adverse effects of drought and soil salinity on crops.

As a new technique combining drip irrigation with film mulching, the mulched drip irrigation has been successfully applied in arid and semi-arid regions in China for more than 20 years [9]. The application area of mulched drip irrigation in Xinjiang has exceeded 2.0×10^6 ha, achieving remarkable economic and social benefits [10]. Drip irrigation keeps the soil in the crop root system moist and the soil in the crop rows relatively dry. This 'dry–wet' interface has a positive effect on regulating soil salinity change and redistributing water and salt. Moreover, mulched drip irrigation can improve soil structure, control temperature and humidity, and then affect the distribution of soil water and salt. Therefore, mulched drip irrigation has been widely used as one of the most suitable irrigation techniques for saline-alkali land [11–13]. Studies have shown that the soil in the cotton root zone generally shows desalination under mulched drip irrigation, but the depth of the desalted zone are different [10,14,15]. There are different views on the change and distribution of soil salinity with the years of mulched drip irrigation. Some studies show that under long-term mulched drip irrigation in cotton fields, the soil salinity in the 0–40 cm or 0–60 cm soil layer decreases year by year, and in some shallow groundwater areas even the whole unsaturated zone is desalted [10,16–18]. Other study suggests that under mulched drip irrigation, the soil within the film is in a state of desalination during the growth period of cotton, while after the growth period, soil salinity in the 0–60 cm layer increases, and the soil salinity increases as a whole with the continuous application of mulched drip irrigation [15]. Meng et al. [19] found that soil salinity decreased in the first three years of mulched drip irrigation, and then increased. Sampling and tracking observation on farmland under mulched drip irrigation were carried out for 5 years, which showed that soil salt did not accumulate significantly [20]. In addition, the distribution of soil salinity in arid areas is affected by many factors, such as groundwater level [21], groundwater salinity [22], irrigation system [11,23], and soil texture [24,25].

In general, previous research results indicate that mulched drip irrigation can provide the necessary soil and water environment for crop growth on saline-alkali land, which is beneficial to increase crop yield. Compared with surface irrigation, mulched drip irrigation has a significant water-saving effect, and plays a positive role in guiding the regulation of soil water and salt and improving saline-alkali land in different areas. However, a consistent conclusion about the long-term effects of mulched drip irrigation on soil salinity has not formed, and even contradictory conclusions were drawn in some studies. Thus, it is of great significance for agricultural development in arid and semi-arid regions to study the dynamics and distribution of soil salinity under long-term mulched drip irrigation.

This paper presents the change characteristics of soil salinity under long-term drip irrigation and the impact of soil texture and soil water on soil salinity, in order to provide a basis for soil salt management under long-term mulched drip irrigation in arid and semi-arid areas. The main objectives of this study were to: (i) investigate interannual changes and the characteristics of soil salinity change during the growth period in salt wasteland under long-term mulched drip irrigation; (ii) analyze the vertical distribution of soil salinity and its relationship with soil texture; (iii) characterize soil water and salinity distribution at different distances from the dripper.

2. Materials and Methods

2.1. Experimental Sites

The experimental sites are located in the irrigated areas of the Shihezi reclamation area in Xinjiang Province (Figure 1). The climate of the study area is typical arid [26]. The average annual precipitation and potential evaporation are 148 and 1900 mm respectively. The groundwater depth is 2–4.5 m, and groundwater mineralization is generally greater than 10 g L^{-1}. Extreme weather, scarce rainfall and strong evaporation make an efficient water-saving irrigation technique the only way to sustainably develop the oasis.

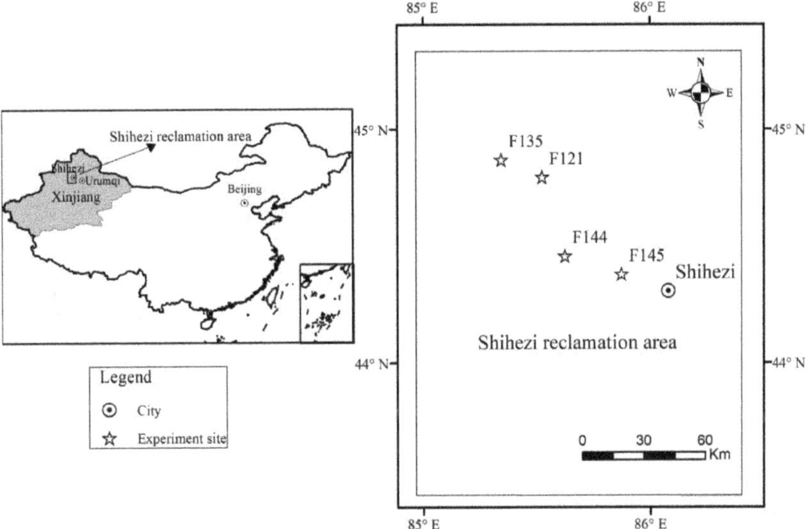

Figure 1. Location map of experimental sites in Shihezi reclamation area, Xinjiang, China.

2.2. Experimental Design and Field Management

The experiment was carried out on four farms, and one newly reclaimed salt wasteland was selected from each of the four farms (the numbers of the farms are F121, F135, F144, and F145; Figure 1) in the experimental area for long-term salt dynamic observation. Each field, with an area of 2000 m^2, was newly reclaimed in 2008 using mulched drip irrigation. In 2016, an experimental field that had been under mulched drip irrigation for 8 years was selected from farm F121 to study the distribution of soil water and salt in the transect perpendicular to the drip pipe. The entire field experiment was conducted from 2008 to 2016. The crops planted were medium long-staple cotton (Xinluzao series); the maximum embedded depth of taproot exceeds 1 m, while most roots are mainly distributed in the 0–60 cm soil layer. According to the climatic characteristics, the cotton was usually sown in the middle of April (dry sowing wet germination) and picked in the middle of September after about 150 days. During the intermission, agricultural operations such as tillage, soil loosening, and pesticide spraying were carried out, and cotton topping was conducted in mid-July. The planting mode adopted the 'one pipe, one film, and four rows' cotton arrangement method. The inner diameter of drip line is 16 mm, the spacing between drip holes is 300 mm, and the designed emitter flow rate is 2.6 L h^{-1}. The width of plastic film was 110 cm, the row spacing of cotton plants was 20 cm, and cotton plants on both sides were 23 cm away from the drip pipe, as shown in Figure 2.

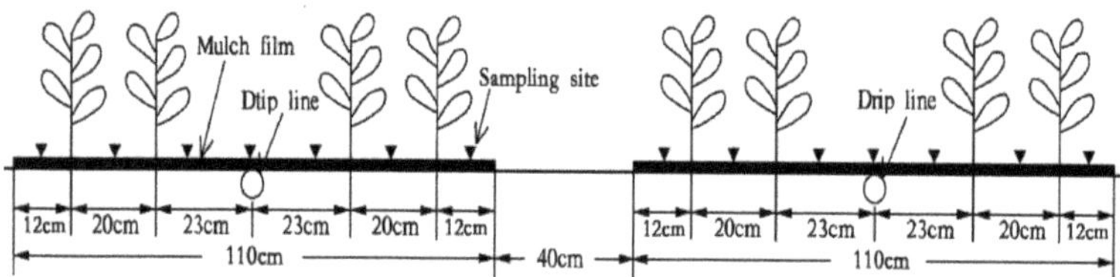

Figure 2. Schematic diagrams of cotton planting, mulching pattern, dripper location, and sampling sites.

Irrigation and fertilization systems were in accordance with local production practices: Irrigating water 8–10 times during the growth period with an irrigation quota of 732–789 mm, and the farmland idle without irrigation treatment during the non-growth period. The water source for irrigation comes from the Manasi River and the salinity of the river water is 0.50~0.97 g L^{-1}. According to the growth of cotton, different fertilizers, mainly urea and potassium ammonium phosphate, were used for each irrigation. The annual amount of fertilizer applied was about 1041 and 708 kg ha^{-1}, respectively. Only shallow tillage and land leveling were applied to the cotton field in the non-growth period.

2.3. Data Collection

For salt wasteland sampling, three sampling areas were selected for each plot using a diagonal sampling method, and soil samples were collected in and out of the mulch film, respectively, in each sampling area. The sampling depths were 0–10, 10–20, 20–30, 30–40, 40–60, and 60–100 cm. From 2011, Soil samples from each layer were analyzed to determine their texture. The sampling was carried out before irrigation in April (pre-sowing), May (seedling stage), June (bud stage), July (blossoming and boll forming stage), August (boll opening stage), and September (harvest period) in 2008–2016.

To study the distribution of soil water and salinity in the transect perpendicular to the drip pipe, sampling work was arranged at seven different sites (Figure 2) along the direction perpendicular to drip irrigation pipe in F121 on 20 June, 23 June, 30 June, and 8 July 2016. In particular, during this period, a single irrigation was carried out on June 21, and the sampling depth of each sampling point were 0–10, 10–20, 20–30, 30–40, 40–60, and 60–100 cm.

Soil water content was determined using an oven-drying method and was given on a mass basis. After measuring the soil water content, soil salinity was measured using weight method [27]. This method is based on water extract from soil sample. The extract is dried to constant weight. We used 15% H_2O_2 to remove the organic matter in the residue. What remained were the total water-soluble salts from the soil. Detailed procedures for soil salinity measurements are presented in the reference. The soil salinity and soil water content in this paper are the average value of multiple samples in each soil layer. The soil salinity is expressed by weight (g kg^{-1}) and the soil water content is expressed in percent of weight (%).

2.4. Statistical Analysis

Soil salinity dynamics and statistical analysis were performed using EXCEL 2010 (Microsoft Corp., Redmond, Washington, DC, USA) and Origin 9.0 (Origin Lab Corp., Northampton, MA, USA). Spatial distribution of soil water and salinity was presented through typical contours using Origin 9.0.

3. Results and Discussion

3.1. Interannual Variation Characteristics of Soil Salinity under Mulched Drip Irrigation during 2008–2016

The changes of soil salinity in the 0–30 and 0–60 cm layers are shown in Figure 3. The soil salinity in the 0–30 cm and 0–60 cm layers is the average value of the corresponding soil layers. As shown in Figure 3a, remarkable decline stage of soil salinity exists in the first few years whether in the 0–30 or 0–60 cm layer with an increase in the years of mulched drip irrigation. The trend lines (Figure 3) indicate that soil salinity decreased sharply from 2008 to 2011, and soil salinity was well correlated with time. Of those, soil salinity in F121 showed the biggest drop. The average soil salinity of F121 in the 0–30 cm layer decreased from 41.8 g kg^{-1} in 2008 to 5.4 g kg^{-1} in 2011, and the average soil salinity in the 0–60 cm layer decreased from 33.7 g kg^{-1} in 2008 to 4.8 g kg^{-1} in 2011. With 2011 as a dividing line of time, the soil decreasing rates in the two periods (before and after 2011) are shown in Table 1. From 2008 to 2011, the 0–30 cm and 0–60 cm soil layers presented a state of desalination with a mean decreasing rate of 6.20 g kg^{-1} per year in 0–30 cm layer and 4.39 g kg^{-1} per year in 0–60 cm layer. While after 2011, soil salinity showed a slight upward trend with a mean decreasing rate of −0.19 g kg^{-1} per year in 0–30 cm layer and −0.3 g kg^{-1} per year in 0–60 cm layer. It is worth noting that the soil salinity of the four fields as a whole shows an obvious downward trend before 2011 or 2012, and a slight upward fluctuation after 2011 or 2012. In other words, the effect of soil desalination is obvious when drip irrigation is applied for 3 years (up to 2011) or 4 years (up to 2012). After that, soil resalinization seems to occur. This result is consistent with previous research results that soil salinity decreases significantly in the initial stage of applying mulched drip irrigation [19].

Table 1. Decrease rate of soil salinity in different soil layers in the salt wastelands, 2008–2016.

Depth (cm)	Time Period	Decrease Rate of Soil Salinity (g kg^{-1} Year^{-1})				Mean (%)
		F121	F135	F144	F145	
0–30	2008–2011	13.44	2.32	2.21	6.81	6.20
	2011–2016	−0.11	−0.21	−0.03	−0.40	−0.19
0–60	2008–2011	10.20	1.42	2.50	3.44	4.39
	2011–2016	−0.14	−0.45	−0.31	−0.49	−0.30

Soil salt seems to re-accumulate after 3–4 years of mulched drip irrigation. The nature of irrigation water may be an influencing factor. Meng et al. [19] pointed out that the amount of salt ions brought into soil through irrigation water (with a salinity of 0.4 g L^{-1}) was 1247 kg ha^{-1} per year, accounting for 5.6% of the initial total soil salt content. Therefore, it could be believed that salt carried by irrigation water is related to soil salt re-accumulation under long-term mulched drip irrigation. In addition, the groundwater in the study area is relatively shallow (2–4.5 m), hence the groundwater brings salt into the soil through evaporation and capillarity under the condition of no drip irrigation and film mulching in the non-growth period, resulting in soil resalinization [22,28]. Moreover, the absence of leaching effects from winter and spring irrigation aggravates this phenomenon [21]. Many studies have shown that soil salinity decreases as the years of mulched drip irrigation increase [3,16,17]. In this study, soil salinity decreases obviously only in the first 3–4 years. Differences in experimental plots, irrigation systems, and field management may be the reasons for the different results. Study also shows that under mulched drip irrigation combined with flood irrigation, soil salt will not accumulate significantly [20]. Therefore, in order to ensure soil health and crop growth, there is an urgent necessity to carry out flood irrigation and salt exclusion measures for salt leaching over a time interval (such as 3–4 years), especially in the non-growth periods.

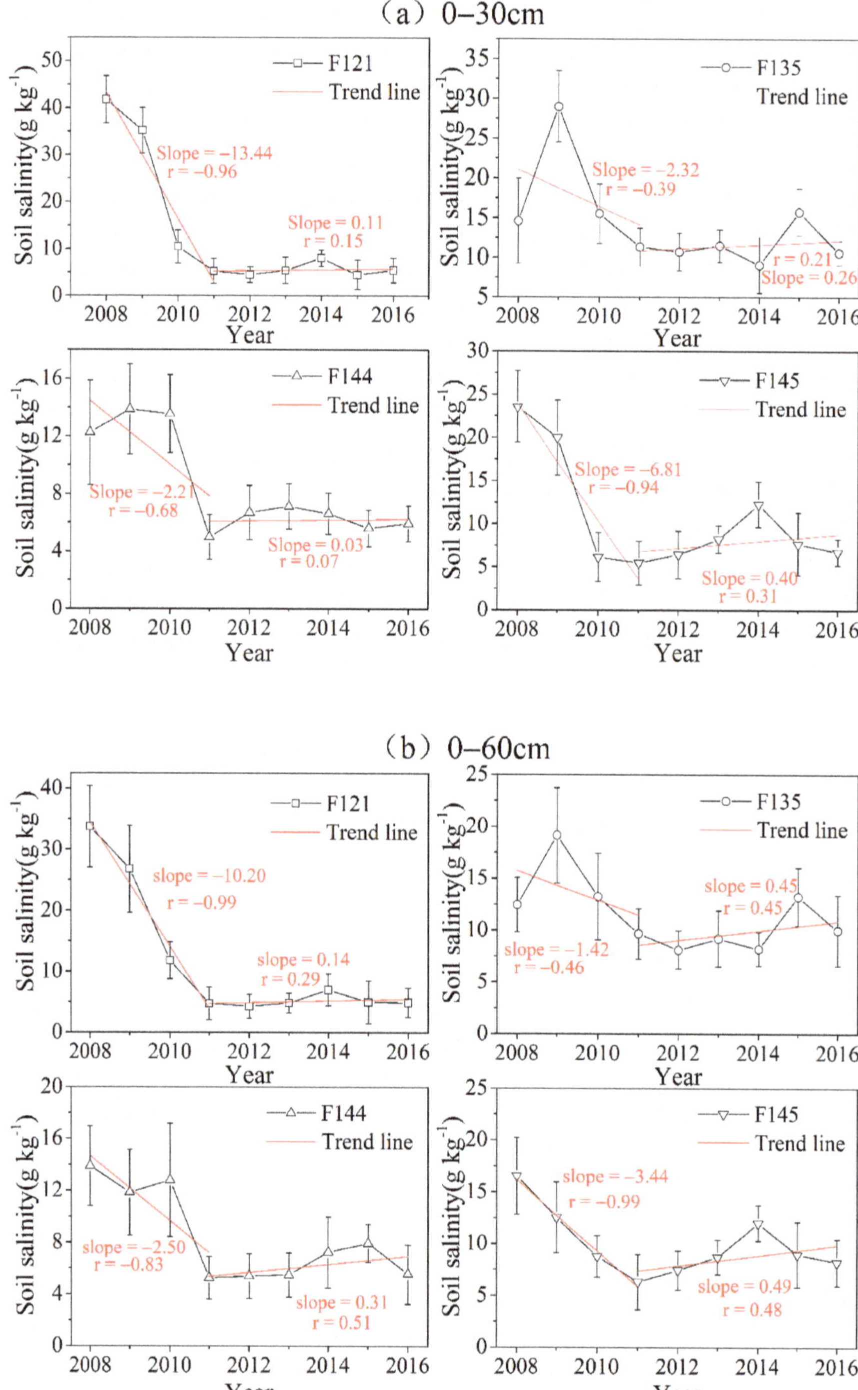

Figure 3. Dynamics of soil salinity in 0–30 cm (**a**) and 0–60 cm (**b**) layers.

3.2. Variation Characteristics of Soil Salinity during Growth Period under Mulched Drip Irrigation

The changes of soil salinity in the 0–60 cm soil layer during the cotton growing periods from 2008 to 2016 are shown in Figure 4. Soil salinity is at a high level at pre-sowing each year. With the beginning of drip irrigation, soil salinity decreases obviously and then fluctuates. During the harvest period in September, soil salinity tends to rise again. This phenomenon has a lot to do with farmland management in the non-growth period. Without irrigation and mulching in the non-growth period, soil salt cannot be leached. During the non-growth period, the groundwater level gradually restores and reaches its peak in the following spring [22], and the salt in groundwater reaches the upper soil layer through evaporation and capillarity [29]. Immediately after irrigation, soil salt is transported to the deep layer due to leaching of irrigation water. During the harvest period in September, soil salt cannot be leached with the cessation of irrigation, and the salt in the deep layer begins to move upward [30]. Therefore, the control of groundwater level and leaching of soil salt through flood irrigation in the non-growth period seem to be important measures to regulate irrigation-induced soil salinity and provide a healthy soil environment for cotton growth in growth period.

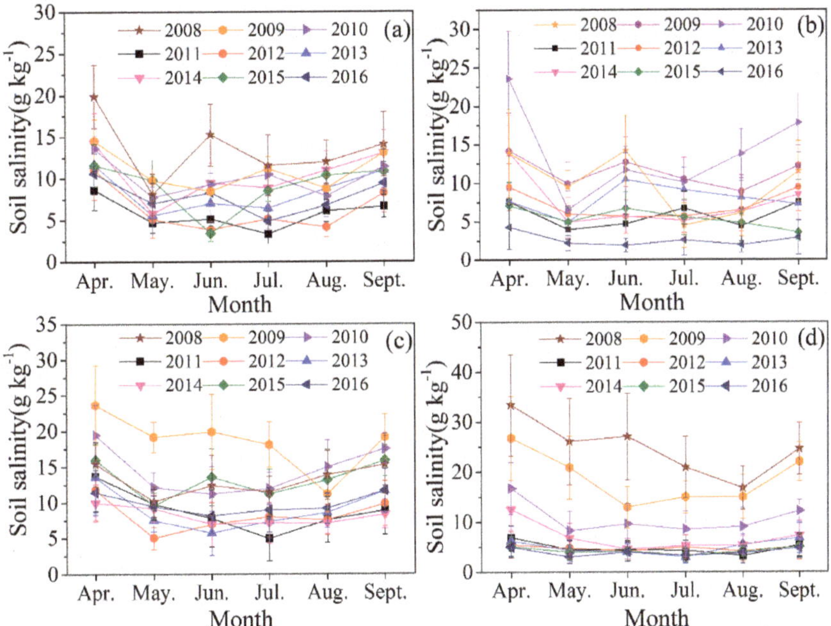

Figure 4. Changes of soil salinity in the 0–60 cm soil layer during the growth period in (**a**) F121, (**b**) F135, (**c**) F144, and (**d**) F145.

3.3. Soil Salinity Distribution along the Vertical Direction in Different Years

The distribution of soil salinity at different depths in 2011–2016 is shown in Figure 5, which also shows the soil texture of each layer. As depicted in Figure 5, the soil salinity of F121 basically decreases with depth in all years except for the relatively high soil salinity in the 60–100 cm layers. According to the soil texture of different soil layers in F121, the soil texture is loam above 30 cm and sandy soil below 30 cm. Compared with loam, sandy soil, with a larger particle size, has stronger permeability, which is more conducive to the movement and diffusion of soil water and salt with the infiltration of irrigation water [31]. Therefore, the soil salinity in the upper loam is higher than that in the bottom sandy soil. Similarly, since the layered soil texture of F135 is very similar to that of F121, the soil salinity of F135 also decreases with an increase of depth. In contrary, the soil salinity of F144 increases with depth as a whole due to the soil texture is loam in the upper part and heavy loam in the bottom part. Notably, the soil salinity of F144 accumulates in the soil layer transiting from loam to heavy loam. Also, the soil texture of F145 is loam in the upper part and clay in the lower part, and the soil salt accumulates in the soil layer transiting from loam to clay. Based on the soil texture, there is a common feature of soil salinity distribution in the vertical direction, that is, soil salt accumulates near the soil layer transiting from coarse soil to fine soil.

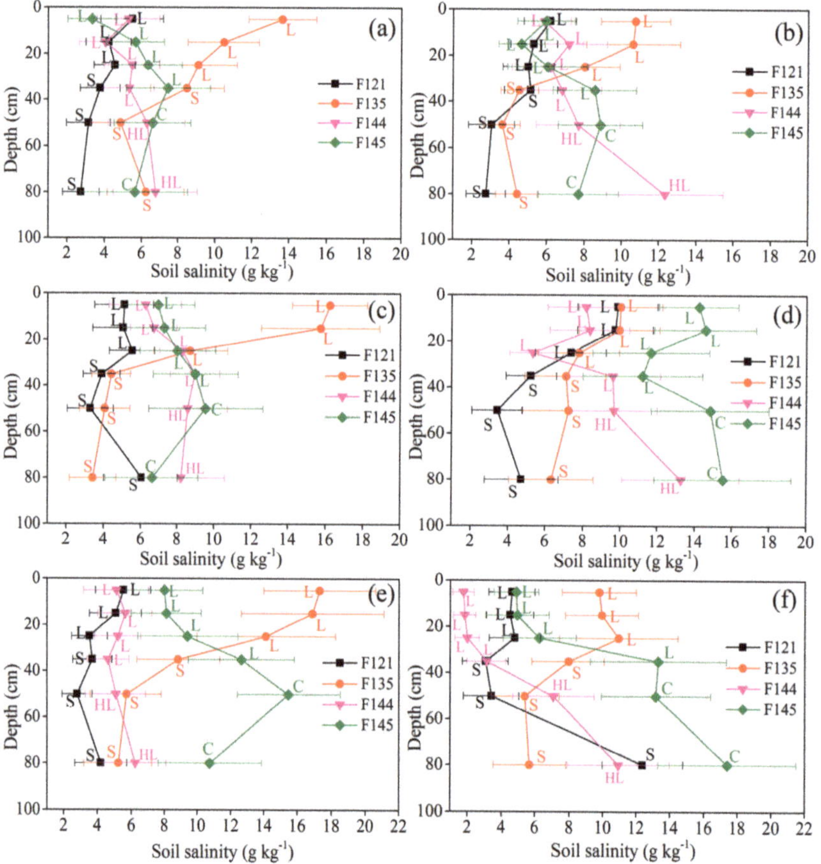

Figure 5. Distribution of soil salinity and soil texture at different soil depths in salt wastelands under mulched drip irrigation, 2011–2016 ((**a**)–(**f**): 2011–2016, S: sand, L: loam, HL: heavy loam, C: clay).

The permeability of the fine soil particle layer is poor, restricting the infiltration of irrigation water. Moreover, the fine-grained soil layer prevents the salt in the groundwater from migrating to the shallow coarse soil layer, thus alleviating the salinization of shallow soil caused by groundwater evaporation, especially in the region with shallow groundwater level [25,32]. In addition, compared with phreatic evaporation, downward infiltration of irrigation water is dominant under mulched drip irrigation. Therefore, soil salt will accumulate near the fine soil layer due to the relatively higher permeability of coarse soil than fine soil. However, for the soil layer with fine particles in the upper part and coarse particles in the bottom part, the leaching effect of irrigation water in the upper soil layer is poor, so soil salt will accumulate in the upper part, while the soil permeability in the bottom part is good, and it is easy for soil salt to be leached to the deep layer. Generally, the poor soil water and salt movement caused by uneven vertical distribution of soil texture leads to soil salt accumulation. Zhang et al. also found that soil salt would accumulate above the relatively impermeable layer [25]. Therefore, soil texture has a great influence on the distribution of soil salinity. Deep tillage can improve water infiltration, facilitate soil water storage and water retention [33–35], reduce soil bulk density and promote salt migration to deep layer, especially for continuous cropping farmlands [35]. Therefore, deep tillage of the farmland should be adopted regularly to alleviate salt accumulation in the root zone.

3.4. Effects of Soil Water Distribution on Soil Salt Distribution before and after a Single Irrigation

3.4.1. Soil Water Distribution in the Soil Profile Perpendicular to the Drip Line

The distribution of soil moisture in the vertical profile is shown in Figure 6. Before irrigation, in the horizontal direction, the soil water content near the dripper is the lowest, and slightly increases in the direction away from the dripper. The reason may be that soil water move to cotton roots under the action of root water uptake. In the vertical direction, the soil water content below the dripper is lower than that at other locations within the 0–30 cm layer. The soil water content in the layer below 30 cm is markedly increased, all more than 20%. In the 30–60 cm layer under the cotton plants, there are wet bulbs with high soil water content (Figure 6a). The reason might be that the cotton roots absorb water from the surrounding zone.

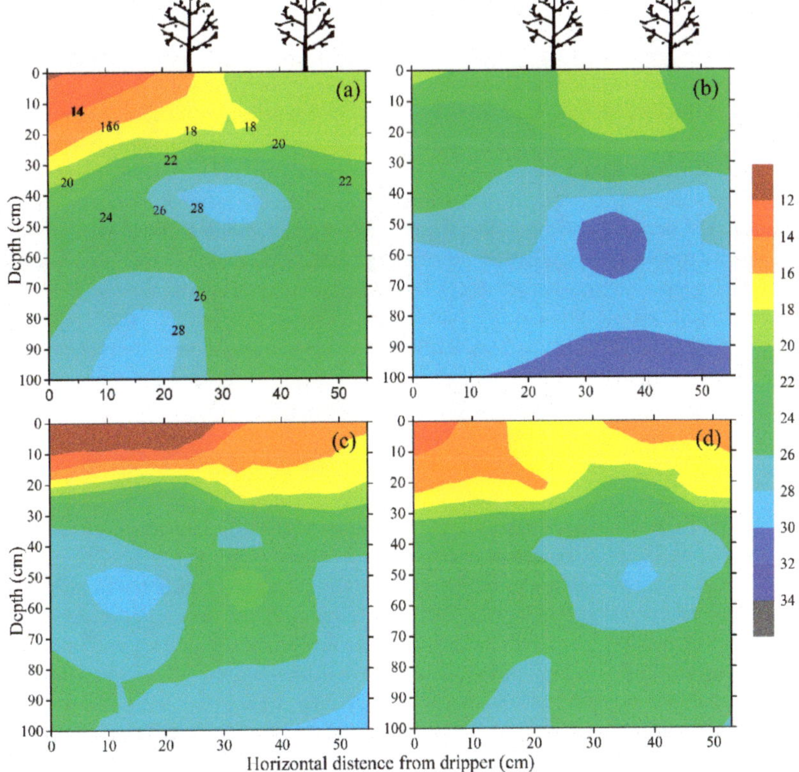

Figure 6. Vertical distribution of soil water content (%, by mass) under mulched drip irrigation on (**a**) 20 June, (**b**) 23 June, (**c**) 30 June, (**d**) 8 July 2016.

Following a single irrigation, the soil water content increases appreciably and presents a nearly stratified distribution after a period of infiltration (Figure 6b). The slight fluctuation in the same layer might be related to the cotton root and soil microstructure, which affects soil permeability [36]. Similarly, there are wet bulbs with high soil water content in the soil layer (40–70 cm) where the roots of cotton plants are relatively concentrated. A few days after irrigation (Figure 6c), the soil water content in each layer decreases, among which the soil water content of the top 40 cm layer decreases obviously; in the shallow soil layer, the farther away from the dripper, the higher the soil water content. Notably, in the 40–70 cm layer under cotton plants, the soil water content is lower than that in the surrounding zone. The decrease of soil water content in shallow soil is related to the continuous downward infiltration of irrigation water, while the decrease of soil water at 40–70 cm depth under cotton plants may due to absorption of water in the process of root growth.

Until the next irrigation (Figure 6d), the soil water in the shallow soil layer increases slightly, while there are obvious wet bulbs at the depth of 40–70 cm under the cotton plants. This phenomenon indicates that the water infiltration driven by drip irrigation is very weak. However, cotton is at the flowering and boll setting stages, during which water consumption is the greatest [37]. Therefore, in the absence of continuous irrigation water, cotton roots not only absorb water from the surrounding zones, but also consume the soil moisture generated from phreatic evaporation under intense evapotranspiration to meet the needs of cotton growth.

In general, before and after an irrigation activity, the soil water content near the dripper first increases and then decreases. This should be due to the constant infiltration of irrigation water. However, the change of soil water content in the root zone is more complex. The distribution of soil water is closely related to crop roots [38]. After irrigation, the well-developed cotton root system in the root zone will absorb water quickly, and the surrounding water moves towards the root system under the action of water potential gradient to supply sufficient water for the root system and make it develop better. Therefore, soil water content in the root zone will increase following irrigation. After a period of infiltration, the zones with developed root system in turn have a stronger ability to absorb water, so that the soil water content in the root zone was relatively lower than that in surrounding zones (Figure 6c). When irrigation stopped, the infiltration of irrigation water is gradually replaced by the upward migration of groundwater caused by phreatic evaporation, crop evapotranspiration and capillarity. The upward gaseous water condenses and accumulates in shallow soil, resulting in a slight increase in soil water content in the shallow layer. The wet bulbs in the root zone (Figure 6d) are due to the cotton root water uptake, especially in periods of high water-consumption. Moreover, there are great differences in root system distribution in the different growth stages of cotton [39]. These differences will result in differences in root water uptake [38], leading to differences in soil water content distribution with horizontal location and soil depth [40].

3.4.2. Soil Salinity Distribution in the Soil Profile Perpendicular to the Drip Line

Distribution of soil salinity in soil profile perpendicular to the drip line before and after one drip irrigation is shown in Figure 7. Before irrigation (Figure 7a), the closer to the dripper, the lower the soil salinity is in the shallow soil layer. The high salinity zone is mainly concentrated in 50–70 cm soil layer under the cotton plants. Following drip irrigation (Figure 7b), the desalinization zones in the soil layer above 70 cm obviously enlarge, and approximately show quarter annuluses distribution with the dripper as the center. While the high salinity zones move down, and the salt accumulates in the soil layer below 70 cm under the cotton plants. From Figure 7c, it can be seen that the soil salinity decreases in the 0–20 cm layer under the dripper but increases slightly in the 30–70 cm layer. While under the cotton plants, soil salinity increases slightly in the 20–60 cm layer and greatly in the soil layer below 60cm. This might be due to that the weakening of irrigation water's downward infiltration enhances the upward movement of soil water in the deep layer under the strong effect of evapotranspiration. Therefore, soil salts move upward with the water mobilization. Comparatively, soil salinity in the 0–20 cm soil layer changes little in the whole soil profile on July 8 (Figure 7d), while soil salinity increases greatly in layers below 20 cm under the cotton plants. Likewise, the upward movement of salt with groundwater caused by the lack of irrigation water infiltration and crop evapotranspiration might be the main reason.

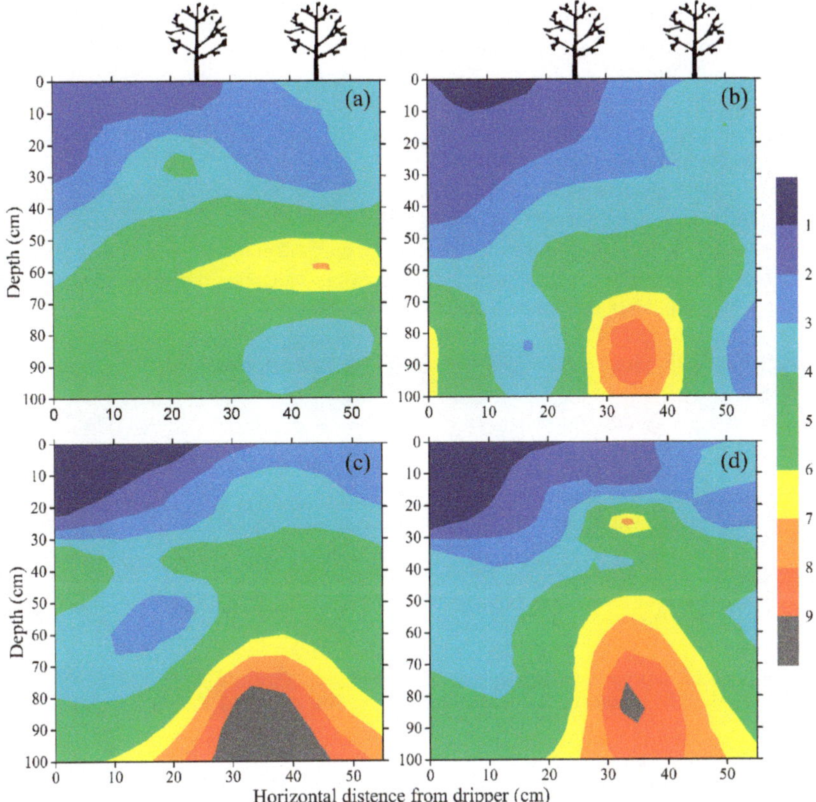

Figure 7. Vertical distribution of soil salinity (g kg^{-1}) under mulched drip irrigation on (**a**) 20 June, (**b**) 23 June, (**c**) 30 June, (**d**) 8 July 2016.

To sum up, after a single irrigation, soil salinity in the soil layer above 70 cm decreases, and soil salt accumulates in the soil layer below 70 cm under the cotton plants. Several days after irrigation, the soil salinity in the shallow layer do not change dramatically, while under the cotton plant, soil salinity shows a rising trend in the soil layer below 30 cm. There is a close relationship between soil water distribution and salt distribution under mulched drip irrigation [41]. Therefore, the distribution and change of soil salinity should be analyzed from the perspective of soil water. In the initial stage after irrigation, the infiltration of soil water from top to bottom takes a dominant position, and soil salt moves down with soil water, resulting in the decrease of soil salinity in the shallow layer and the accumulation of soil salt in the deep soil layer, especially under cotton plants [42]. After a period of time, the cessation of irrigation leads to the weakening of water infiltration, while upward groundwater migration is dominant due to phreatic evaporation, capillarity and crop transpiration [43,44]. As a result, the salt in the groundwater gradually moves upward and accumulates as shown in Figure 7. These changes in soil salinity after irrigation are consistent with the study of Qi et al., which showed that the soil salt tended to move upward from deeper layer to top layer as time passed after irrigation [41]. Soil salinity shows an upward trend before the next irrigation. Therefore, the irrigation interval should be shortened, and the irrigation frequency should be appropriately increased to leach and restrain soil salt. Irrigation is leaching process for soil salt, and a suitable irrigation schedule is of great significance for soil desalination under mulched drip irrigation.

4. Conclusions

In this study, the soil salinity of newly reclaimed salt wasteland was monitored from 2008–2016 based on field experiments. Besides that, the distribution of soil salinity and water in the transect perpendicular to the drip pipe before and after a single drip irrigation was also investigated from 20 June 2016 to 8 July 2016. From the results obtained in this work, the following can be concluded that:

Under long-term mulched drip irrigation, the soil salinity in 0–30 cm and 0–60 cm layers showed a sharp decline in the first 3 to 4 years and then began to fluctuate and showed an upward trend. During the growth period, soil salinity was generally higher at pre-sowing and late harvest period, and decreased immediately after drip irrigation. Soil texture and soil water seriously affect the dynamics and distribution of soil salinity. Soil salt will accumulate in the soil layer transiting from coarse to fine soil from top to bottom. After a single irrigation, soil salt will migrate first downward and then upward with the change of soil water. Therefore, corresponding measures such as flood irrigation, deep tillage, optimization of irrigation regime, and salt exclusion hydraulic measures should be applied to alleviate soil resalination and promote the development of agriculture under long-term mulched drip irrigation. The quantitative study on the effects of these measures on soil desalination under long-term mulched drip irrigation will be the topics in our future research.

Author Contributions: Conceptualization, Z.G. and Z.J.; Methodology, Z.G. and Z.Z.; Investigation, Z.G., Z.J., Z.Z. and Q.Y.; Resources, Z.J., Z.Z. and Q.Y.; Writing—Original Draft Preparation, Z.G.; Writing—Review and Editing, Z.G. and Z.J.

Funding: This work was founded by the '111' Project (B08039), National Natural Science Foundation of China (41877179), Special Fund for Basic Scientific Research of Central Colleges (300102299206), Open Fund of Key Laboratory of Subsurface Hydrology and Ecological Effect in Arid Region of Ministry of Education (300102298502), and China Postdoctoral Science Foundation (2018M633438).

Acknowledgments: The authors would like to thank all members at Paotai Soil Improvement Test Station for monitoring and providing places and facilities in the experiments, and thank Proof-Reading-Service.com for language help.

Conflicts of Interest: The authors declare no conflict of interests.

References

1. Rozema, J.; Flowers, T. Crops for a salinized world. *Science* **2008**, *322*, 1478–1480. [CrossRef]
2. Yang, S.; Zhang, F.; Liu, S. A New Method for Reforming Salt Soil-Genetic Transformation with Genes Tolerance to Salt. *World For. Res.* **2006**, *19*, 14–19.
3. Li, W.; Wang, Z.; Zheng, X.; Zhang, J. Soil salinity variation characteristics of cotton field under long-term mulched drip irrigation. *Trans. Chin. Soc. Agric. Eng.* **2016**, *32*, 67–74. (In Chinese)
4. Ayers, R.S.; Westcot, D.W. *Water Quality for Agriculture*; Food and Agriculture Organization of the United Nations: Rome, Italy, 1985; Volume 29.
5. Seydehmet, J.; Lv, G.; Nurmemet, I.; Aishan, T.; Abliz, A.; Sawut, M.; Abliz, A.; Eziz, M. Model Prediction of Secondary Soil Salinization in the Keriya Oasis, Northwest China. *Sustainability* **2018**, *10*, 656. [CrossRef]
6. Peck, A.J.; Hatton, T. Salinity and the discharge of salts from catchments in Australia. *J. Hydrol.* **2003**, *272*, 191–202. [CrossRef]
7. Dregne, H.E. Land Degradation in the Drylands. *Arid Land Res. Manag.* **2002**, *16*, 99–132. [CrossRef]
8. Houk, E.; Frasier, M.; Schuck, E. The agricultural impacts of irrigation induced waterlogging and soil salinity in the Arkansas Basin. *Agric. Water Manag.* **2006**, *85*, 175–183. [CrossRef]
9. Xu, F.; Li, Y.; Ren, S. Investigation and discussion of drip irrigation under mulch in Xinjiang Uygur Autonomous Region. *Trans. Chin. Soc. Agric. Eng.* **2003**, *19*, 25–27. (In Chinese)
10. Li, M.; Liu, H.; Zheng, X. Spatiotemporal variation for soil salinity of field landunder long-term mulched drip irrigation. *Trans. Chin. Soc. Agric. Eng.* **2012**, *28*, 82–87. (In Chinese)

11. Liu, M.; Yang, J.; Li, X.; Liu, G.; Yu, M.; Wang, J. Distribution and dynamics of soil water and salt under different drip irrigation regimes in northwest China. *Irrig. Sci.* **2013**, *31*, 675–688. [CrossRef]

12. Mmolawa, K.; Or, D. Root zone solute dynamics under drip irrigation: A review. *Plant Soil* **2000**, *222*, 163–190. [CrossRef]

13. Wang, R.; Kang, Y.; Wan, S.; Hu, W.; Liu, S.; Liu, S. Salt distribution and the growth of cotton under different drip irrigation regimes in a saline area. *Agric. Water Manag.* **2011**, *100*, 58–69. [CrossRef]

14. Zhang, W.; Lv, X.; Li, L.; Liu, J.; Sun, Z.; Zhang, X.; Yang, Z. Salt transfer law for cotton field with drip irrigation under the plastic mulch in Xinjiang Region. *Trans. Chin. Soc. Agric. Eng.* **2008**, 15–19. [CrossRef]

15. Mu, H.; Hudan·Tumaerbai; Su, L.; Mahemujiang·Aihemaiti; Wang, Y.; Zhang, J. Salt transfer law for cotton field with drip irrigation under mulch in arid region. *Trans. Chin. Soc. Agric. Eng.* **2011**, *27*, 18–22. (In Chinese)

16. Wang, Z.; Yang, P.; Zheng, X.; He, X.; Zhang, J.; Li, W. Soil salinity changes of root zone and arable in cotton field with drip irrigation under mulch for different years. *Trans. Chin. Soc. of Agric. Eng.* **2014**, *30*, 90–99. (In Chinese)

17. Li, W.; Wang, Z.; Zheng, X.; Zhang, J. The effect of drip irrigation under film in long-term on the soil salinity in root zone and the cotton growth. *J. Arid Land Resour. Environ.* **2015**, *8*, 161–166. (In Chinese)

18. Wang, Z.; Yang, P.; Zheng, X.; He, X.; Zhang, J.; Li, W. Soil Salt Dynamics in Cotton Fields with Mulched Drip Irrigation under the Existing Irrigation System in Xinjiang. *Trans. Chin. Soc. Agric. Mach.* **2014**, *45*, 149–159. (In Chinese)

19. Meng, C.; Yan, L.; Zhang, S.; Wei, C. Variation of Soil Salinity in Plow Layer of Farmlands under Long-term Mulched Drip Irrigation in Arid Region. *Acta Pedol. Sin.* **2017**, *6*, 1386–1394. (In Chinese)

20. Hu, H.; Tian, F.; Zhang, Z.; Yang, P.; Ni, G.; Li, B. Soil salt leaching in non-growth period and salinity dynamics under mulched drip irrigation in arid area. *J. Hydraul. Eng.* **2015**, *46*, 1037–1046. (In Chinese)

21. Ming, G.; Tian, F.; Hu, H. Effect of water table depth on soil water and salt dynamics and soil salt accumulation characteristics under mulched drip irrigation. *Trans. Chin. Soc. Agric. Eng.* **2018**, *34*, 90–97. (In Chinese)

22. Abliz, A.; Tiyip, T.; Ghulam, A.; Halik, Ü.; Ding, J.; Sawut, M.; Zhang, F.; Nurmemet, I.; Abliz, A. Effects of shallow groundwater table and salinity on soil salt dynamics in the Keriya Oasis, Northwestern China. *Environ. Earth Sci.* **2016**, *75*, 260. [CrossRef]

23. Zhang, G.; Liu, C.; Xiao, C.; Xie, R.; Ming, B.; Hou, P.; Liu, G.; Xu, W.; Shen, D.; Wang, K.; et al. Optimizing water use efficiency and economic return of super high yield spring maize under drip irrigation and plastic mulching in arid areas of China. *Field Crop. Res.* **2017**, *211*, 137–146. [CrossRef]

24. Hu, H.; Tian, F.; Hu, H. Soil particle size distribution and its relationship with soil water and salt under mulched drip irrigation in Xinjiang of China. *Sci. China Technol. Sci.* **2011**, *54*, 1568–1574. [CrossRef]

25. Zhang, Z.; Hu, H.; Tian, F.; Hu, H.; Yao, X.; Zhong, R. Soil salt distribution under mulched drip irrigation in an arid area of northwestern China. *J. Arid Environ.* **2014**, *104*, 23–33. [CrossRef]

26. Jia, Z.; Zhao, Z.; Zhang, Q.; Wu, W. Dew Yield and Its Influencing Factors at the Western Edge of Gurbantunggut Desert, China. *Water* **2019**, *11*, 733. [CrossRef]

27. Bado, S.; Forster, B.P.; Ghanim, A.M.A.; Jankowicz-Cieslak, J.; Berthold, G.; Luxiang, L. Protocol for measuring soil salinity. In *Protocols for Pre-Field Screening of Mutants for Salt Tolerance in Rice, Wheat and Barley*; Springer: Cham, Switzerland, 2016.

28. Wang, Q.; Wang, W.; Wang, Z.; Zhang, J.; Zhang, J.; Ding, X.; Zhou, L.; Mao, G. Analysis of features of soil salt in a drainage area. *Acta Pedol. Sin.* **2001**, *38*, 271–276.

29. Wu, J.; Zhao, L.; Huang, J.; Yang, J.; Vincent, B.; Bouarfa, S.; Vidal, A. On the effectiveness of dry drainage in soil salinity control. *Sci. China Series E Technol. Sci.* **2009**, *52*, 3328–3334. [CrossRef]

30. He, Z.; Shi, W.; Yang, J. Water and salt transport and desalination effect of halophytes intercropped cotton field with drip irrigation under film. *Trans. Chin. Soc. Agric. Eng.* **2017**, *33*, 129–138. (In Chinese)

31. Ma, W.; Zhang, X.; Zhen, Q.; Zhang, Y. Effect of soil texture on water infiltration in semiarid reclaimed land. *Water Qual. Res. J. Can.* **2015**, *51*, 33–41. [CrossRef]

32. Yu, S.; Yang, J.; Liu, G. Effect of clay interlayers on soil water-salt movement in easily-salinized regions. *Adv. Water Sci.* **2011**, *22*, 495–501. (In Chinese)

33. Blanco-Canqui, H.; Wienhold, B.J.; Jin, V.L.; Schmer, M.R.; Kibet, L.C. Long-term tillage impact on soil hydraulic properties. *Soil Tillage Res.* **2017**, *170*, 38–42. [CrossRef]

34. Schneider, F.; Don, A.; Hennings, I.; Schmittmann, O.; Seidel, S.J. The effect of deep tillage on crop yield—What do we really know? *Soil Tillage Res.* **2017**, *174*, 193–204. [CrossRef]

35. Cui, J.; Tian, L.; Guo, R.; Lin, T.; Xu, H.; Li, F. Effect of Deep Tilling on Soil Moisture Content and Salinity Content of Drip Irrigation Cotton Continuous Cropping. *Chin. Agric. Sci. Bull.* **2014**, *30*, 134–139. (In Chinese)

36. Ranaivomanana, H.; Razakamanantsoa, A.; Amiri, O. Permeability Prediction of Soils Including Degree of Compaction and Microstructure. *Int. J. Geomech.* **2017**, *17*, 04016107. [CrossRef]

37. Wang, M.; Yang, Q.; Zheng, J.; Liu, Z. Spatial and temporal distribution of water requirement of cotton in Xinjiang from 1963 to 2012. *Acta Ecol. Sin.* **2016**, *13*, 4122–4130. (In Chinese)

38. Coelho, E.F.; Or, D. Root distribution and water uptake patterns of corn under surface and subsurface drip irrigation. *Plant Soil* **1998**, *206*, 123–136. [CrossRef]

39. Mai, W.; Tian, C.; Li, C. Soil Salinity Dynamics under Drip Irrigation and Mulch Film and Their Effects on Cotton Root Length. *Commun. Soil Sci. Plan* **2013**, *44*, 1489–1502. [CrossRef]

40. Li, X.; Šimůnek, J.; Shi, H.; Yan, J.; Peng, Z.; Gong, X. Spatial distribution of soil water, soil temperature, and plant roots in a drip-irrigated intercropping field with plastic mulch. *Eur. J. Agron.* **2017**, *83*, 47–56. [CrossRef]

41. Qi, Z.; Feng, H.; Zhao, Y.; Zhang, T.; Yang, A.; Zhang, Z. Spatial distribution and simulation of soil moisture and salinity under mulched drip irrigation combined with tillage in an arid saline irrigation district, northwest China. *Agric. Water Manag.* **2018**, *201*, 219–231. [CrossRef]

42. Chen, W.; Jin, M.; Ferre, T.P.A.; Liu, Y.; Xian, Y.; Shan, T.; Ping, X. Spatial distribution of soil moisture, soil salinity, and root density beneath a cotton field under mulched drip irrigation with brackish and fresh water. *Field Crop. Res.* **2018**, *215*, 207–221. [CrossRef]

43. Nishida, K.; Shiozawa, S. Modeling and Experimental Determination of Salt Accumulation Induced by Root Water Uptake. *Soil Sci. Soc. Am. J.* **2010**, *74*, 774. [CrossRef]

44. Jin, X.; Chen, M.; Fan, Y.; Yan, L.; Wang, F. Effects of Mulched Drip Irrigation on Soil Moisture and Groundwater Recharge in the Xiliao River Plain, China. *Water* **2018**, *10*, 1755. [CrossRef]

 © 2019 by the authors. Licensee MDPI, Basel, Switzerland. This article is an open access article distributed under the terms and conditions of the Creative Commons Attribution (CC BY) license (http://creativecommons.org/licenses/by/4.0/).

Article

Modeling of Fertilizer Transport for Various Fertigation Scenarios under Drip Irrigation

Romysaa Elasbah [1], Tarek Selim [1,*] , Ahmed Mirdan [1] and Ronny Berndtsson [2,3]

[1] Civil Engineering Department, Faculty of Engineering, Port Said University, Port Fouad 42523, Egypt; eng.romysaa@yahoo.com (R.E.); a_mirdan@yahoo.com (A.M.)

[2] Division of Water Resources Engineering, Lund University, Box 118, SE-221 00 Lund, Sweden; ronny.berndtsson@tvrl.lth.se

[3] Center for Middle Eastern Studies, Lund University, Box 201, SE-221 00 Lund, Sweden

* Correspondence: eng_tarek_selim@yahoo.com

Received: 1 April 2019; Accepted: 20 April 2019; Published: 28 April 2019

Abstract: Frequent application of nitrogen fertilizers through irrigation is likely to increase the concentration of nitrate in groundwater. In this study, the HYDRUS-2D/3D model was used to simulate fertilizer movement through the soil under surface (DI) and subsurface drip irrigation (SDI) with 10 and 20 cm emitter depths for tomato growing in three different typical and representative Egyptian soil types, namely sand, loamy sand, and sandy loam. Ammonium, nitrate, phosphorus, and potassium fertilizers were considered during simulation. Laboratory experiments were conducted to estimate the soils' adsorption behavior. The impact of soil hydraulic properties and fertigation strategies on fertilizer distribution and use efficiency were investigated. Results showed that for DI, the percentage of nitrogen accumulated below the zone of maximum root density was 33%, 28%, and 24% for sand, loamy sand, and sandy loam soil, respectively. For SDI with 10 and 20 cm emitter depths, it was 34%, 29%, and 26%, and 44%, 37%, and 35%, respectively. Results showed that shallow emitter depth produced maximum nitrogen use efficiency varying from 27 to 37%, regardless of fertigation strategy. Therefore, subsurface drip irrigation with a shallow emitter depth is recommended for medium-textured soils. Moreover, the study showed that to reduce potential fertilizer leaching, fertilizers should be added at the beginning of irrigation events for SDI and at the end of irrigation events for DI. As nitrate uptake rate and leaching are affected by soil's adsorption, it is important to determine the adsorption coefficient for nitrate before planting, as it will help to precisely assign application rates. This will lead to improve nutrient uptake and minimize potential leaching.

Keywords: drip irrigation; fertilizer transport; fertigation strategy; adsorption coefficient; HYDRUS-2D/3D

1. Introduction

Fertilization practice includes application of nitrogen, phosphate, and potassium before planting. For this purpose, manual broadcasting and mechanical spreading or spraying are used. Fertigation can be defined as the process of mixing irrigation water with fertilizers. Fertigation promotes overall root activity, improves nutrient mobility and uptake, as well as mitigating pollution of surface water and groundwater [1,2]. In Egypt, fertigation is practiced on only 13% of agricultural lands [3]. The fertigation technique is mainly used with nitrogen (N) and potassium (K) fertilizers [4].

For Egyptian agricultural conditions, nitrogen is considered one of the most important factors in crop production. Due to its excessive application by farmers, combined with poor surface irrigation management, N concentration is, on average, 1.50 ppm in the Nile Delta drains [3]. Therefore, sustainable agricultural management should be adopted. This management must include water saving irrigation methods (e.g., drip irrigation) along with precise estimation of fertilizer application rates to mitigate harmful effects of excessive use of fertilizers on the surrounding environment.

The most common N fertilizers used in Egypt are nitrate (NO_3^-), ammonium (NH_4^+), and urea. Nitrate and ammonium are absorbed and used by crops, while urea is hydrolyzed to ammonium by heterotrophic bacteria, then nitrified to nitrite and nitrate by autotrophic bacteria [5]. Nitrate is highly mobile and easily leaches, due to its negative charge. Thus, excessive application of N might lead to nitrate contamination of surface water and groundwater [6]. Potassium, also adsorbs weakly to soil particles. Therefore, intensive use of fertilizers may increase N and K concentrations in groundwater [7,8]. Phosphorus leaching and runoff are insignificant, because phosphorus is usually adsorbed to particles, and is thus considered almost immobile [9].

Irrigation methods, soil hydraulic properties, management practices, climatic parameters, crop type, and crop rotation are major factors affecting risk for fertilizer leaching [10–12]. To achieve maximum fertilizer use efficiency, a proper fertigation strategy associated with modern irrigation technology should be implemented. Drip irrigation is considered a modern irrigation system that provides a great degree of control for both irrigation water and fertigation, allowing accurate application in accordance with crop water requirements and thereby reducing fertilizer leaching. Moreover, it allows for a controlled placement of nutrients near the plant roots, limits fertilizer losses, and reduces fertilizer leaching to the groundwater. Full understanding of water and fertilizer distribution patterns in the root zone and fertilizer leaching below the root zone are required for proper design of drip fertigation systems, regarding application rate and duration [13]. Considering all these parameters through large-scale field experiments is labor-, time-, and cost-consuming. Numerical simulations, on the other hand, are an inexpensive alternative which can help in assessing either existing or proposed fertigation practices.

The HYDRUS-2D/3D model [14] can efficiently simulate two or three dimensional water flow and fertilizer (e.g., urea–ammonium–nitrate, phosphorus, and potassium) transport in partially saturated porous media. In addition, it can simulate root water extraction and root nutrient uptake. Many researchers have shown that HYDRUS-2D/3D is a suitable software with which to simulate water flow and solute transport under different irrigation systems and fertigation scenarios (e.g., References [5,15–24]). Cote et al. [15] used HYDRUS-2D/3D to simulate water and fertilizer movement under subsurface trickle irrigation, considering different fertigation strategies for bare soil. They found that fertigation at the beginning of irrigation cycles can reduce nitrogen leaching. Gardenas et al. [16] evaluated nitrate leaching for various fertigation scenarios under micro irrigation systems, considering different soil types (sandy loam, loam, silt, and clay), assuming that the adsorption coefficient for nitrate was set equal to zero based on the assumption that the soils were free from positive charges, as nitrate is negatively charged, and, if soil is free from positive charges, nitrate will not adsorb to soil particles (i.e., adsorption coefficient for nitrate = 0). This also assumes that no other processes occur that affect the retention or release of soil chemical or biological nitrate content. They demonstrated that seasonal leaching was highest for coarse-textured soil, and that it increased when applying fertilizers at the beginning of the irrigation cycle, as compared to fertigation at the end of irrigation cycle. They showed that nitrate uptake by plant roots is smaller for micro-sprinkler irrigation, as compared to drip irrigation systems. The N use efficiency varied from 20 to 30% in micro-sprinkler irrigation. However, it reached 40 to 60% in drip irrigation. Hanson et al. [5] studied N uptake and leaching using urea–ammonium–nitrate fertilizers for surface drip (DI) and subsurface drip irrigation (SDI) systems. They ignored the adsorption behavior for urea and nitrate. They found that N use efficiency was about 50 to 65% for DI and 44 to 47% for SDI. Ajdari et al. [17] used HYDRUS-2D/3D to investigate nitrate leaching for an experimental onion field under DI, considering different emitter discharges. They found that as the emitter discharge increases, the amount of nitrogen leaching out from the root zone increases. The same conclusion was reached by Shekofteh et al. [25]. In addition, they found that simulation of nitrate movement using HYDRUS 2D/3D was close to the measured results in field soils.

As a result of the lack of irrigation water, some of Egypt's agricultural land suffers from chemical contamination due to illegal practices such as use of untreated drainage water from industry and agriculture for irrigation purposes. Examples of sources of contaminants are organic phosphorus pesticides, chlorinated hydrocarbon pesticides, rodenticides, and a variety of other pesticides including lead arsenate, calcium arsenate, copper oxides, and mercury [26]. Therefore, it is necessary to investigate the effects of soil composition and charge, especially for land areas suffering from chemical contamination, on nitrate adsorption behavior that affect leaching rates and N use efficiency. Nitrate can be adsorbed to soil particles if they contain positive charges. Iron and aluminum oxide concentrations in soil, organic matter content, and soil texture affect the nitrate adsorption rates to soil particles.

The present study introduces DI and SDI as alternatives to traditional irrigation methods (flood and furrow irrigattion) in order to overcome the problems associated with water shortage and to protect the environment from excessive application of nitrogen fertilizers using surface irrigation methods. Consequently, the HYDRUS-2D/3D model was used to investigate fertilizer distribution (i.e., ammonium, nitrate, phosphorus, and potassium) under DI and SDI, considering different fertigation strategies and soil types cultivated with tomato crops. The effect of fertilizer adsorption on the fertilizer uptake rates and fertilizers' leaching was also investigated.

2. Materials and Methods

The HYDRUS-2D/3D model (version 2.04, PC-Progress, Prague, Czech Republic) was used to simulate fertilizer movement under DI and SDI of tomatoes growing in three different soil types (sand, loamy sand, and sandy loam) representing typical Egyptian soils (Typic Xeropsamments to Typic Psammiaquents). Ammonium, nitrate, phosphorus, and potassium fertilizers were considered during simulations. Moreover, the impact of soil hydraulic properties, fertigation strategy, and fertilizer adsorption behavior, such as fertilizer distribution and losses, were investigated. HYDRUS-2D/3D is a computer software package used to simulate water, solute (i.e., chemicals), and heat transport in two or three dimensional variably saturated porous media. The HYDRUS-2D/3D model uses the Galerkin finite element method to solve the modified form of Richards' equation, which includes a sink term to consider water uptake by plant roots for simulating water flow. The model solves the Fickian-based advection–dispersion equation for solute transport (e.g., [27]). The transport equations contain terms for nonlinear non-equilibrium reactions between the solid and liquid phases and two first-order degradation reactions. For more details of the HYDRUS code and its application, see Reference [14].

2.1. System Description

In the DI setup, the spacing between drip lines was set to 140 cm (one drip line per plant row), and the spacing between emitters was set to 35 cm with an emitter flow rate of $1\ L\ h^{-1}$. The simulation domain used was rectangular with a 100 cm soil depth and a 70 cm width, with a tomato plant located at the upper left corner of the simulation domain. The SDI system was arranged to have the same characteristics as the DI system with emitter depths of 10 and 20 cm below the soil surface (Figure 1). Unstructured triangular mesh with 21,086 and 21,896 2D elements was used to spatially discretize the flow domain for the DI and SDI, respectively, with smaller size mesh elements at the surface and close to the emitter.

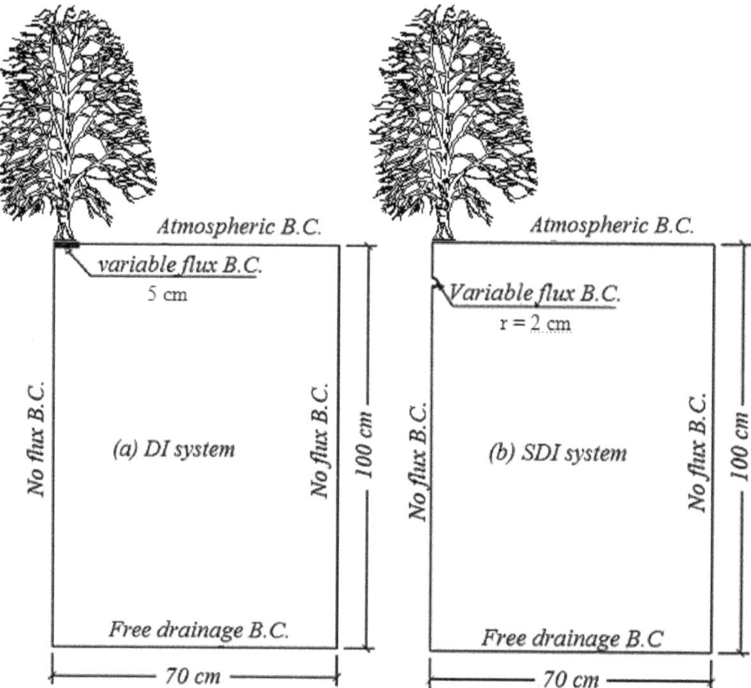

Figure 1. Simulation domain for the investigated drip irrigation (DI) and subsurface drip irrigation (SDI) systems.

Soil Parameters

Three different soil types (sand, loamy sand, and sandy loam) representing three typical and representative agricultural soil types in Egypt were considered. Table 1 shows the physical properties of the three different soil types.

Table 1. Physical properties for the different soil types used in the simulations (Typic Xeropsamments to Typic Psammiaquents).

Soil Type	Sand (%)	Silt (%)	Clay (%)	Bulk Density (gm/cm^3)
Sand	92.0	3.6	4.4	1.66
Loamy sand	79.7	12.3	8.0	1.73
Sandy loam	65.2	30.8	4.0	1.78

Soil hydraulic parameters were taken according to Abou Lila et al. [28], as the same fields were considered in the present study. Table 2 shows the soil hydraulic parameters used during model simulation, and Table 3 summarizes observed chemical parameters for the simulated soil types. Table 3 shows that all simulated soils have positive charges, due to the existence of heavy metals and other contaminants. These positive charges will affect the adsorption behavior of nitrate and other fertilizers.

Table 2. Hydraulic parameters for the different soil types used in the simulations.

Soil Type	θ_r (m^3 m^{-3})	θ_s (m^3 m^{-3})	α	n	K_s (cm day^{-1})	l
Sand	0.024	0.447	0.124	1.87	878.20	0.5
Loamy sand	0.074	0.453	0.045	1.72	288.50	0.5
Sandy loam	0.038	0.486	0.025	1.72	194.06	0.5

θ_r: residual water content. θ_s: saturated water content. K_s: saturated hydraulic conductivity. α: inverse of the air-entry value. n: pore size distribution index, l: pore connectivity parameter.

Table 3. Observed chemical characteristics for soil types used in the simulations.

Soil Type	pH	Calcium Carbonate (%)	Na (meq/L)	Mg (meq/L)	Ca (meq/L)	Fe (ppm)	Cu (ppm)	Zn (ppm)
Sand	8.5	0.7	1.5	1.2	1.1	6.9	0.86	1.9
Loamy sand	8.0	2.1	2.7	11.1	8.2	5.5	0.24	3.0
Sandy loam	7.9	0.8	40.1	12.4	11.1	12.8	0.74	2.8

2.2. Solute Parameters

Molecular diffusion coefficient (D_w), first-order decay coefficients (μ_w, μ_s), and adsorption isotherm coefficient (K_d) for each type of fertilizer were required for simulation implementation. Longitudinal and lateral dispersivities (ε_L and ε_t, respectively) were also required. Thus, ε_L was set = 0.1 m (i.e., one tenth of the simulation domain [29,30], while ε_t was set = 0.1 $\times$ ε_L [30]. Molecular diffusion was set equal to 1.957 $\times$ 10^{-5} cm^2 s^{-1} for NH$_4^+$ and K$^+$ [31], and 1.902 $\times$ 10^{-5} cm^2 s^{-1} for NO$_3^-$ [25,31].

The rate coefficients of nitrification from ammonium to nitrate (μ_w, μ_s) were set equal to 0.2 day^{-1} [32,33].

2.3. Batch Experiments

Batch experiments were carried out to estimate fertilizer adsorption isotherm parameters for the different soil types. The batch experiments were conducted for each type of fertilizer and repeated for each type of soil, as described by Flury and Fluhler [34]. Soil samples were sieved through a 2 mm sieve and dried over one day at 105 °C. Solutions with initial concentrations (C_o) of 124, 60, 46.5, and 132.3 ppm of potassium, phosphorus, ammonium, and nitrate were prepared using potassium sulfate, phosphoric acid, ammonium sulfate, and calcium nitrate, respectively. A 25 mL of the prepared solution of each fertilizer was mixed with 25 g soil and shaken in an inert Teflon flask for 3 h at 20 °C. After that, soil and solution were separated by 30 min centrifugation and the concentration of fertilizers in the supernatant solution was measured. Flame photometer was used to measure potassium concentration, while spectrophotometer at a wavelength of 660 nm was used to estimate phosphorus concentration and Kjeldahl to measure nitrogen concentration. The adsorbed mass of fertilizer (C_a; mg kg^{-1}) was calculated based on mass balance (the difference compared to the total mass of fertilizers). The mass found in the liquid phase at equilibrium (C_l; g m^{-3}) was assumed to be adsorbed by the soil. The experiments were made in duplicate.

Adsorption isotherm parameters were calculated using linear Freundlich isotherm equations ((Equation (1)).

$$C_a = K_d C_l \tag{1}$$

where K_d is the distribution coefficient (dm^3 kg^{-1}). Values of the distribution coefficient for different types of fertilizer in the different soil types used in the model simulations are presented in Table 4.

Table 4. Distribution coefficient for fertilizers in the different soil types (cm^3 g^{-1}).

Soil Type	Sand	Loamy Sand	Sandy Loam
Potassium	2.50	3.26	3.99
Phosphorus	3.62	4.31	4.41
Ammonium	2.20	2.50	2.86
Nitrate	0.65	0.72	0.89

In contrast to most previous research that assumed K_d for nitrate to be equal to zero, the batch experiments gave an input for the K_d for all collected soil samples. Most previous research (e.g., [5,16]) has assumed that nitrate is not adsorbed to soil particles with negative charge. However, in our case, the chemical analyses of collected soil samples (Table 3) revealed the presence of heavy metals and other contaminants that have positive charges. It is worth mentioning that the collected soil samples for this study were taken from the El-Salam Canal region. The El-Salam Canal water is considered brackish, as it is a mixture of Nile water and salty agricultural drainage water. It may also contain industrial waste water.

The volatilization of ammonium and subsequent ammonium transport by gaseous diffusion were neglected during simulations.

2.4. Initial Conditions

The initial soil water content (θ_i) in soil was assumed to be uniform in the entire flow domain. The effective saturation (θ_e) was set equal to 0.25 m^3 m^{-3} for all soil types, in order to determine θ_i according to:

$$\theta_e = \frac{\theta - \theta_r}{\theta_s - \theta_r} \tag{2}$$

where θ is the volumetric water content equal to θ_i at $t = 0$. The resulting θ_i values were equal to 0.13, 0.15, and 0.169 m^3 m^{-3} for sand, sandy loam, and loamy sand, respectively. The simulation domain was assumed to be free of fertilizers at the beginning of simulations.

2.5. Boundary Conditions

Figure 1 shows the imposed boundary conditions (B.C.) assumed during simulation of DI and SDI. No flux B.C. was set along vertical sides of the simulation domain. The left side was set as zero flux B.C. due to symmetry, and as the result of using a large flow domain the right boundary was set to zero flux B.C. as well. The bottom boundary was assigned as free drainage B.C., as the groundwater table is situated 1.50 m below the soil surface. In SDI, the top boundary was set as the atmospheric B.C. that allows for crop evapotranspiration (ET$_c$) along the whole length of the upper boundary of the simulation domain. In DI, the location of the emitter was set as variable flux B.C., and the remaining part of the top boundary was assigned as atmospheric B.C. The ET$_c$ value was taken as a constant (0.75 cm day^{-1}), as calculated by Selim et al. [35] using the same study area. Although the HYDRUS-2D/3D model required partitioning of ET$_c$ to evaporation (E) and transpiration (T), T was set equal to ET$_c$, while E was assumed to be zero during the simulation period. T was set equal to ET$_c$, as the simulation was conducted only during the mid-growth season of the tomato crop, for which the surface area of land was approximately covered by tomato leaves (i.e., crop canopy). A constant flux (q) of 68.57 and 109.14 cm day^{-1} was used at the emitter location during irrigation events in the case of DI (Equation (3)) and SDI (Equation (4)), respectively. When irrigation was terminated, these fluxes were converted to no flux boundary condition.

$$q = \frac{\text{Emitter discharge flow rate}}{\text{Drip tubing surface area}} = \frac{1 \times 1000 \times 24}{10 \times 35} = 68.57 \text{ cm day}^{-1} \tag{3}$$

where the flux diameter (10 cm) was assumed to be equal to the wetted diameter to avoid numerical simulation instability.

$$q = \frac{\text{Emitter discharge flow rate}}{\text{Drip tubing surface area}} = \frac{Q}{2\,\pi\,r\,S} = \frac{1 \times 1000 \times 24}{2 \times \pi \times 1 \times 35} = 109.14 \text{ cm day}^{-1} \tag{4}$$

where Q is the emitter discharge ($L^3\ T^{-1}$), r is the emitter radius (L), and S is the distance between emitters (L).

As fertilizers were assumed to be applied with irrigation water, third-type Cauchy B.C. was set at the top edge of the simulation domain and along the emitter location for both DI and SDI.

2.6. Fertilizer Application

Fertilizers were added to the tomato crop with irrigation water according to the agricultural bulletin for tomato issued by the Egyptian Ministry of Agriculture. Four fertigation strategies were assumed in this study. The irrigation event period was divided into three equal intervals. In the first fertigation strategy (strategy B), fertilizers were applied at the beginning of the irrigation period (i.e., during the first third of irrigation). In the second and third fertigation strategies (strategies M and E, respectively), fertilizers were applied at the middle and at the end of the irrigation event period, respectively. In fertigation strategy C, fertilizers were applied during the whole period of the irrigation event. The total amounts of fertilizers added in the entire simulation period were 300.0, 21.4, 128.6, and 200.0 kg/ha for potassium, phosphorus, ammonium, and nitrate, respectively. Phosphorus was only added during the first 21 days, while other fertilizers were added during the entire simulation period. Potassium and phosphorus were added three times a week, while ammonium and nitrate were added twice a week.

2.7. Root Parameters

As the HYDRUS-2D/3D model version 2.04 does not consider root growth, and as a result of the lack of information about the root distribution through the entire growing season of the tomato crop, only simulation of the mid-growth season of tomato crop was executed. The growing stage was selected as the leaf area index for tomato crop is relatively constant, which leads to a constant root-to-shoot ratio [36]. The Vrugt et al. [37] model was used to describe root parameters. The following parameters of Vrugt's model were used as input for the HYDRUS-2D/3D model: Z_m = 100 cm, X_m = 70 cm, z^* = 25 cm, x^* = 0, P_z = 1, and P_x = 1 [5]. The effect of water stress on root water uptake was considered using a threshold water stress response function presented by Feddes et al. [38] with the following parameters: P_o = −1 cm, P_{opt} = −2 cm, P_{2H} = −800 cm, P_{2L} = −1500 cm, P_3 = −8000, r_{2H} = 0.10 cm day^{-1}, and r_{2L} = 0.10 cm day^{-1}.

2.8. Simulation Scenarios

Simulations were conducted to investigate the effect of irrigation method, soil hydraulic properties, fertilizers' adsorption behavior, and fertigation strategy on fertilizer transport when growing tomato crops. Surface and subsurface drip irrigation with emitters at depths 10 and 20 cm below soil surface were considered in the simulation. The simulations were conducted for sand, loamy sand, and sandy loam during a 40 day period representing the mid-growth stage of tomato crop, and irrigation was applied every alternate day with a duration of 7.35 h for each irrigation event. Four different fertilizer types, namely ammonium, nitrate, phosphorus, and potassium, were considered during the current work. Four fertigation strategies were investigated. Strategies B, M, and E lasted 2.45 h, and strategy C lasted 7.35 h. This led to 36 simulation scenarios.

3. Results and Discussion

3.1. Distribution of Fertilizers

Figure 2 visualizes the progress of the fertilizer distribution at an observation point located on the top left corner of the simulation domain for DI, considering fertigation strategy C with different soil types (a: sand, b: loamy sand, and c: sandy loam). It can be observed that the fertilizer concentration increased at the end of fertigation events and then decreased between irrigation events, due to root uptake and adsorption to soil particles. For irrigation events free of fertilizers, fertilizer concentration decreased at the observation point, due to the movement of fertilizers with irrigation water. Potassium concentration increased at the end of the first fertigation event ($t = 0.31$ day), and then it decreased after ceasing irrigation. The same trend occurred throughout the simulation period, but with a significant decline in potassium concentration at 6.31, 20.31, and 34.31 days. These represent irrigation events without fertilizers, as potassium was applied three times a week. Phosphorus followed the same trend as potassium during the first 21 days of simulation. It then decreased with time until the end of simulation, as phosphorus was only applied during the first 3 weeks of simulation. For nitrate, the concentration increased at the end of the first fertigation events ($t = 0.31$ days). After that, nitrate concentration increased due to the nitrification of ammonium to nitrate and negligible denitrification, as the soil was not saturated between irrigation events. As nitrate was applied twice a week, nitrate concentration decreased during the second irrigation event. By applying the second fertigation, associated with the third irrigation event, nitrate concentration increased, but did not reach the concentration it had after the first fertigation event, due to adsorption, root uptake, and movement with irrigation water. Maximum nitrate concentration occurred at $t = 16$ days, due to the application of two subsequent fertigation events. After that and until the end of the simulation period, nitrate concentration decreased and increased but did not reach the maximum value. Ammonium concentration, on the other hand, increased by the end of fertigation events and decreased between fertigation events, due to nitrification to nitrate, adsorption, and root uptake. It is worth mentioning that approximately the same trend occurred in other fertigation strategies. The concentration of fertilizers around the emitter was the lowest for strategy B as compared to other strategies, due to the movement of fertilizers with irrigation water, while it was the highest for strategy E, as a result of high soil moisture content before fertigation. At the end of the simulation, fertilizer concentrations at the observation point for fertigation strategy C were 82.4, 1.1, 21.4, and 88.3 mg/L for sand; 79.7, 1.2, 20.3, 87.4 mg/L for loamy sand; and 76.5, 1.31, 18.9, and 85.5 mg/L for sandy loam for potassium, phosphorus, ammonium, and nitrate, respectively. For fertigation strategy B, they were 49.3, 1.1, 9.3, and 48.7 mg/L for sand; 50.4, 1.2, 9.1, and 50.4 mg/L for loamy sand; and 51.1, 1.3, 8.9, and 52.8 mg/L for sandy loam, respectively. For fertigation strategy M, they were 64.7, 1.1, 14.4, and 66.9 mg/L for sand; 63.7, 1.2, 13.6, and 67.1 mg/L for loamy sand; and 62.8, 1.3, 12.9, and 68.0 mg/L for sandy loam, respectively. However, for fertigation strategy E, they were 133.1, 1.2, 40.6, and 149.3 mg/L for sand; 124.9, 1.3, 38.8, and 144.6 mg/L for loamy sand; and 115.6, 1.4, 35.0, and 135.8 mg/L for sandy loam, respectively. It is worth mentioning that, in order to save space, only the aforementioned results are discussed herein.

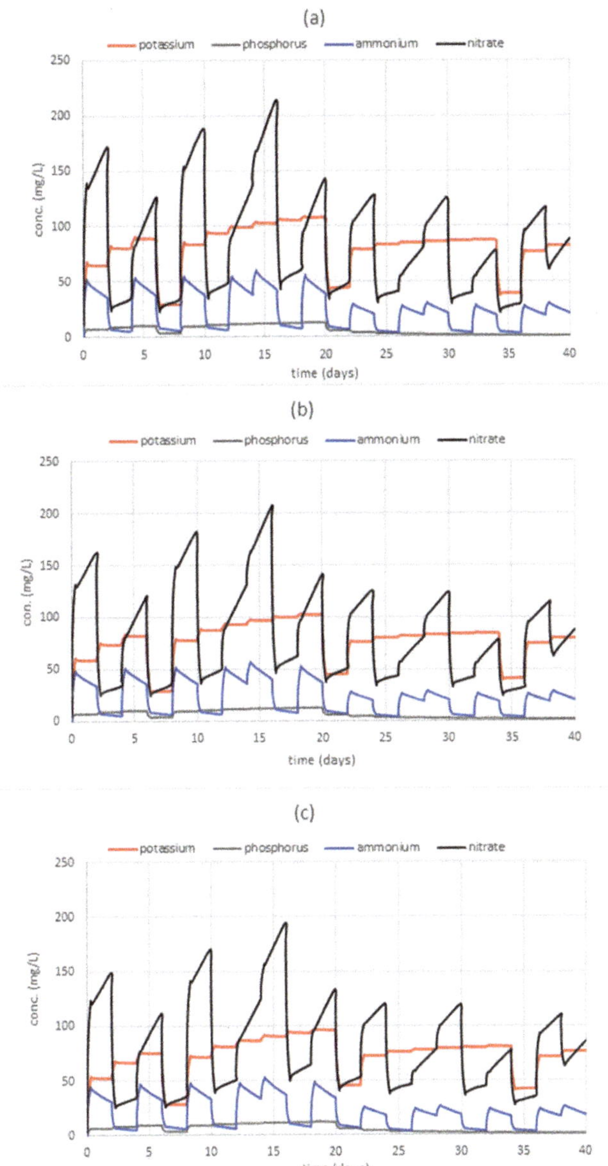

Figure 2. Temporal variation in fertilizer concentration at observation points on the emitter for strategy C using a DI system and different soil types, (**a**) sand, (**b**) loamy sand, and (**c**) sandy loam.

3.2. Effect of Soil Type

Figure 3 illustrates fertilizer distribution in the soil domain after the first fertigation event and at the end of the simulation period for strategy C, considering the three soil types and DI. Soil type affected the fertilizer distribution within the simulation domain. Potassium, phosphorus, ammonium, and nitrate reached soil depths of 48, 38, and 32 cm, 37, 30, and 26 cm, 15, 12, and 10 cm, and 68, 58, and 50 cm below soil surface in sand, loamy sand, and sandy loam, respectively. The figure shows that the fertilizers moved deepest in sandy soil as compared to other soil types. This is because of the low field capacity of sand as compared to other soil types. The lateral spreading in sand, loamy sand, and sandy loam was 28, 30, and 30 cm, 20, 21, and 22 cm, 10, 11, and 11 cm, and 38, 41, and 41 cm for potassium, phosphorus, ammonium, and nitrate, respectively. The downward vertical extent of fertilizers was larger than the lateral extent for all soil types. This can be attributed to the gravity force that dominated during solute transport movement. Lateral movement of fertilizers in loamy sand and sandy loam was higher as compared to sand. This is due to the limited infiltration capacity in fine-textured soil as compared to coarse-textured soils, which led to less air-filled pore space. The adsorption behavior was larger in fine-textured soil as compared to coarse-textured soil. Similar results were obtained under different strategies and SDI (results not shown). It is pertinent to mention that the amount of fertilizers above the emitter in sandy loam soil was higher than for other soil types with SDI. This may be due to the capillary action that increases the upward movement of water and fertilizers in sandy loam as compared to sand and loamy sand soils.

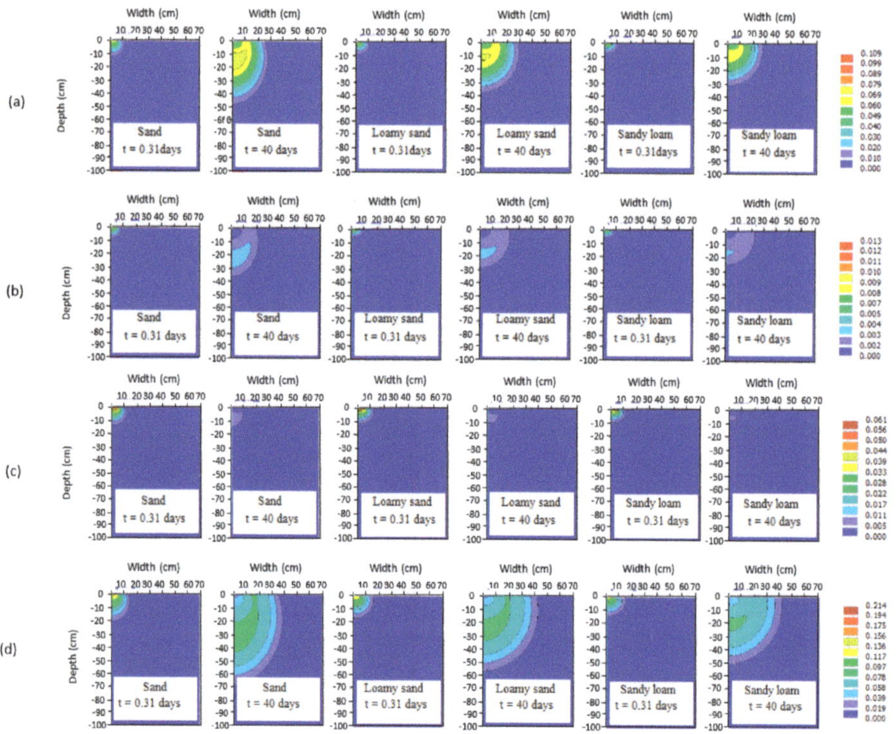

Figure 3. Fertilizer distribution after the first fertigation event ($t = 0.31$ days) and at the end of simulation period ($t = 40$ days) for strategy C for different soil types with DI ((**a**): potassium, (**b**): phosphorus, (**c**): ammonium, and (**d**): nitrate, units: mg cm^{-3}).

3.3. Effect of Irrigation System

Figure 4 shows fertilizer distribution in sandy soil at the end of the simulation period for strategy C, using DI and SDI systems with emitters at 10 and 20 cm depths. It is noted that the fertilizer distribution depends mainly on the location of the emitter. For the DI, potassium, phosphorus, ammonium, and nitrate reached 48, 37, 15, and 68 cm depth below soil surface, respectively. In SDI with an emitter depth of 10 cm, potassium, phosphorus, ammonium, and nitrate reached depths of 50, 41, 24, and 68 cm, respectively. In SDI with an emitter depth of 20 cm, potassium, phosphorus, ammonium, and nitrate moved down to depths of 59, 49, 32, and 77 cm depths. As expected, the downward movement of fertilizers increased as the emitter depth increased. Shallow emitter depths allowed the fertilizers to reach the soil surface and spread more horizontally as compared to the deeper emitters. Thus, large emitter depth may increase the potential risk of groundwater contamination as well as decreasing fertilizer uptake. Similar results were obtained for the other strategies and soil types (results not shown).

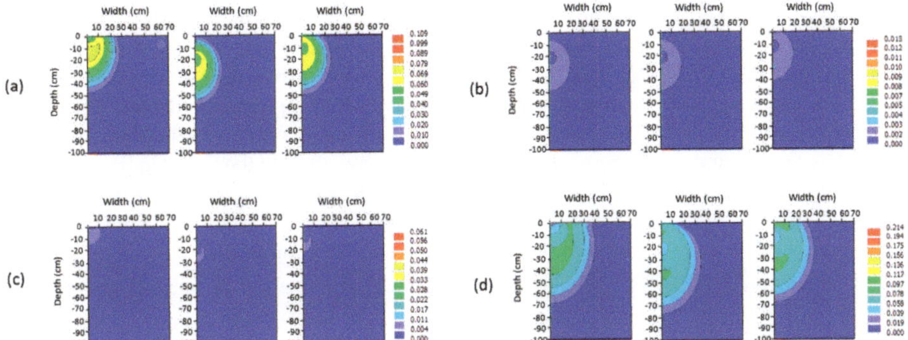

Figure 4. Fertilizer distribution at the end of the simulation period ($t = 40$ days) for strategy C in sand with DI and SDI systems and emitter at 10 and 20 cm depths ((**a**): potassium, (**b**): phosphorus, (**c**): ammonium, and (**d**): nitrate, unit: mg cm^{-3}).

3.4. Fertilizer Leaching

Results of all simulation scenarios showed that nitrate was adsorbed in all soil types under all fertigation scenarios. Therefore, there was insignificant leaching of nitrate outside the soil domain, with a leaching percentage below 1% for all fertigation strategies. No leaching of potassium, phosphorus, and ammonium took place due to high adsorption rates. Results of the amount of fertilizers leaving the simulation domain showed that SDI systems with a shallow emitter depth had the lowest leaching percentage as compared to DI and SDI systems with deep emitter depths. This is attributed to the fact that the fertilizers were applied in the zone of maximum root density, as the emitter was located close to this zone. Thereby, fertilizers could effectively be taken up by plant roots. The results of leaching (accumulated) percentages below the maximum root density (25 cm from the soil surface) are shown in Table 5.

Table 5. Percentage of potassium, phosphorus, and nitrogen accumulated below the maximum root density zone as a fraction of the total potassium, phosphorus, and nitrogen added, respectively.

Soil Type	Fertigation Strategy	DI			SDI with Emitter Depth of 10 cm			SDI with Emitter Depth of 20 cm		
		K	P	N	K	P	N	K	P	N
Sand	C	24.4	26.4	32.7	32.7	36.2	34.0	49.98	52.61	43.30
	E	23.7	25.8	31.9	32.6	36.2	34.2	50.05	52.73	43.34
	M	24.6	26.6	33.0	33.6	37.0	35.0	51.23	53.49	44.41
	B	24.9	26.8	33.1	31.9	35.3	32.9	49.3	51.6	42.2
Loamy sand	C	13.8	15.1	27.7	22.1	25.1	28.7	40.2	42.8	36.3
	E	13.3	14.6	27.2	22.2	25.4	28.9	40.1	43.2	36.5
	M	13.8	15.1	27.8	22.5	25.6	29.1	40.7	43.1	36.7
	B	14.2	15.5	28.2	21.9	24.5	27.7	39.9	42.2	35.7
Sandy loam	C	7.7	10.1	24.4	15.9	20.1	26.3	35.7	39.0	35.4
	E	7.4	9.8	23.9	16.0	20.4	26.7	35.6	39.3	35.5
	M	7.7	10.2	24.4	16.1	20.3	26.6	35.9	39.1	35.5
	B	8.0	10.5	24.9	15.6	19.7	25.6	35.9	38.9	35.1

The table shows that the potential leaching of fertilizers in sand soil is much greater than for loamy sand and sandy loam with DI. For the SDI, the cumulative fertilizers below the maximum root density zone were the lowest for sandy loam soil, due to the upward movement of water and fertilizers that occurred by capillary action. The limited infiltration capacity and the high adsorption characteristics of fine-textured soil particles increase the soil retention of water and fertilizers, making fine-textured soils less susceptible to leaching losses. It is noted that fertigation strategy had little effect on fertilizer leaching. For DI, the potential leaching was the lowest for strategy E and the highest for strategy B. However, for SDI, the leaching potential was the lowest for strategy B. These results concur with the results of Hanson et al. [5] and Cote et al. [15]. They found that fertigation strategy affects leaching to a small degree only. Ajdary et al. [17] found that fertigation strategy did not affect nitrogen leaching, except in the case of coarse-textured soil when applying fertilizers directly before ceasing of the irrigation event. In our study, gravity force dominated fertilizer movement in strategy E, where fertilizers entered the wetting zone and moved downward with the flow. On the other hand, capillarity dominated the movement of fertilizers in strategy B, where the fertilizers moved with water downward and upward by capillary action. Therefore, more fertilizers can be maintained near and above the emitter. Thus, strategy E is recommended to reduce the groundwater contamination risk for DI, and strategy B is recommended for SDI.

3.5. Root Fertilizer Uptake

Table 6 shows that slight differences in the amount of root fertilizer uptake (fertilizer use efficiency) occurred when comparing different fertigation strategies. In our study, fertilizer use efficiency (FUE) is defined as the ratio between the amounts of fertilizer uptake by plant roots to the total amount of fertilizer added to the simulation domain. For the DI, FUE varied in the three soil types from 9.0 to 15.5%, 11.0 to 15.4%, and 26.2 to 35.8% for potassium, phosphorus, and nitrogen, respectively. The largest FUE occurred in sand soil, while the lowest FUE occurred in sandy loam. The higher value of FUE in loamy sand soil as compared to the sandy loam soil may be attributed to higher fertilizer adsorption on sandy loam particles. For SDI with an emitter depth of 10 cm, FUE varied in the three soil types from 9.3 to 15.6%, 11.4 to 15.7%, and 27.2 to 36.4% for potassium, phosphorus, and nitrogen, respectively. For the SDI with an emitter depth of 20 cm, the potassium uptake varied from 9.0 to 14.6%, phosphorus uptake varied from 11.0 to 14.4%, and nitrogen uptake varied from 26.7 to 34.5% in the three soil types. The results show that there is an insignificant difference between root fertilizer uptakes under different fertigation strategies. Consequently, the fertigation strategy does not seem to have any effect on the fertilizer uptake by the plant roots. This result concurs with Gardenas et al. [16]. They reported that nitrate taken up by plant roots was independent of fertigation strategy. However, Hanson et al. [5] found that the best nitrogen uptake ratio occurred when using strategy E for DI, while fertigation strategy did not have any effect on nitrogen uptake for SDI. Results also showed that root nutrient uptake was higher for SDI with a shallow emitter depth as compared to other systems. This may be due to the fact that the emitter was located approximately in the middle of the zone of maximum root density.

This study did not include considerations of practical network effects such as mainline and manifold hydraulics, lag time, and mixing of chemical flow with motive flow (main flow of irrigation water in the lines). However, we still believe that our results display some general and universal results for DI and SDI irrigation system outlines. However, future studies would need to quantify the above network effects.

Table 6. Percentage of root uptake for potassium, phosphorus, and nitrogen as a fraction of the total potassium, phosphorus, and nitrogen added, respectively, for different fertigation strategy.

Soil Type	Fertigation Strategy	DI			SDI with Emitter Depth of 10 cm			SDI with Emitter Depth of 20 cm		
		K	P	N	K	P	N	K	P	N
Sand	C	15.4	15.3	35.5	15.5	15.4	36.2	14.5	14.4	34.3
	E	15.5	15.4	35.8	15.6	15.7	36.4	14.6	14.4	34.5
	M	15.4	15.3	35.5	15.5	15.3	36.0	14.4	14.3	34.0
	B	15.4	15.3	35.2	15.4	15.3	36.2	14.5	14.3	34.5
Loamy sand	C	12.0	12.7	31.9	12.3	13.0	33.1	11.8	12.4	32.2
	E	12.0	12.7	32.1	12.3	13.1	33.3	11.8	12.4	32.2
	M	12.0	12.7	32.0	12.3	13.0	33.0	11.7	12.4	32.2
	B	12.0	12.7	31.7	12.3	13.0	32.9	11.7	12.4	32.0
Sandy loam	C	9.0	11.0	26.3	9.4	11.4	27.4	9.0	11.0	26.8
	E	9.1	11.1	26.5	9.4	11.5	27.6	9.1	11.0	26.8
	M	9.0	11.0	26.3	9.4	11.4	27.3	9.0	11.0	26.7
	B	9.0	11.0	26.2	9.3	11.4	27.2	9.0	11.0	26.7

4. Conclusions

The present study investigated the effect of soil type, fertigation strategy, and adsorption behavior on fertilizer distribution, fertilizer uptake by plant roots, and the amount of fertilizers that can leach below the simulation domain and below the zone of maximum root density for DI and SDI with 10 and 20 cm emitter depths for tomato plants. Simulation results showed that fertilizer leaching is significantly affected by the soil type. Sandy soils were more susceptible to the risk of fertilizer leaching below the maximum root density zone than loamy sand and sandy loam soils. Fertilizer accumulation in sandy loam was also larger than for other soil types. In addition, fertilizer leaching below the simulation domain was affected by varying irrigation systems. SDI with shallow emitter depths had a lower amount of leaching as compared to other drip irrigation systems where water and fertilizers were effectively injected into the zone of maximum root density. Therefore, the amount of fertilizer uptake by roots was the highest for the SDI with a shallow emitter depth. Consequently, a shallow emitter is recommended as compared to deep emitters for SDI, as it reduces the potential risk of groundwater contamination, especially in sandy soil and plants with shallow roots.

Simulation results showed that it is best to conduct fertigation at the end of the irrigation event (strategy E) in the case of DI, and fertigation at the beginning of the irrigation event (strategy B) in SDI. Simulation results showed that nitrate adsorption behavior has a considerable impact on leaching, uptake by plants, and distribution within the soil domain. Logically, as the adsorption coefficient increases, the amount of solute leaching from the soil domain decreases. Additionally, as the emitter discharge and/or the amount of solute increases, the amount of solute leaching from the soil domain increases. Shekofteh et al. [25] used different emitter discharges, varying from 0.5 to 8 L h^{-1}, and different amounts of potassium nitrate, varying from 950 to 2550 kg ha^{-1}, while neglecting the adsorption coefficient of nitrate. They found that, as the emitter discharge and fertilizer increased, the amount of nitrate leaching increased. In any case, it is important to determine the adsorption coefficient for nitrate before planting, as it will help to precisely assign nitrate application rates (fertilizer application rate and duration). This will lead to improved nutrient uptake and minimal leaching.

Author Contributions: Conceptualization, T.S. and A.M.; Methodology, R.E. and T.S.; Software, R.E. and T.S.; Validation, R.E. and T.S.; Formal Analysis, R.E.; T.S.; and R.B.; Investigation, R.E. and T.S.; Resources, T.S. and A.M.; Data Curation, R.E.; T.S. and R.B.; Writing—Original Draft Preparation, R.E. and T.S.; Writing—Review & Editing, R.B.; Visualization, R.E.; T.S. and R.B.; Supervision, T.S. and A.M. Project Administration, T.S. and A.M.; Funding Acquisition, T.S. and A.M.

Funding: This research received no external funding.

Conflicts of Interest: The authors declare no conflict of interest.

References

1. Magen, H. Fertigation: An overview of some practical aspects. *Fert. News* **1995**, *40*, 97–100.
2. Kafkafi, U.; Tarchitzky, J. *A Tool for Efficient Fertilizer and Water Management*; International Fertilizer Association/International Potash institute: Paris/Horgen, France, 2011.
3. Food and Agriculture Organization of the United Nations (FAO). *Fertilizer Use by Crop in Egypt*; Land and Plant Nutrition Management Service, Land and Water Development Division: Roma, Italy, 2005.
4. Taha, M.H. Chemical fertilizers and irrigation system in Egypt. In Proceedings of the FAO Regional Workshop on Guidelines for Efficient Fertilizers Use through Irrigation, Cairo, Egypt, 14–16 December 1999.
5. Hanson, B.R.; Simunek, J.; Hopmans, J.W. Evaluation of urea–ammonium–nitrate fertigation with drip irrigation using numerical modeling. *Agric. Water. Manag.* **2006**, *86*, 102–113. [CrossRef]
6. González-Delgado, A.M.; Shukla, M.K. Transport of nitrate and chloride in variably saturated porous media. *J. Irrig. Drain Eng.* **2014**, *140*, 04014006. [CrossRef]
7. Corley, R.H.V.; Tinker, P.B.H. *The Oil Palm (World Agriculture Series)*, 4th ed.; Wiley-Blackwell: Oxford, UK, 2003; p. 592, ISBN 978-0-632-05212-7.
8. Griffioen, J. Potassium adsorption ratios as an indicator for the fate of agricultural potassium in groundwater. *J. Hydrol.* **2001**, *254*, 244–254. [CrossRef]
9. Feigen, A.; Ravina, I.; Shalhevet, J. Effect of Irrigation with Treated Sewage Effluent on Soil, Plant and Environment. In *Irrigation with Treated Sewage Effluent*; Springer: Berlin, Germany, 1990.
10. Nemeth, T. Nitrogen in the soil-plant system nitrogen balances. *Cer. Res. Commun.* **2006**, *34*, 61–64.
11. Josipovic, M.; Kovačević, V.; Šostarić, J.; Plavšić, H.; Liovic, J. Influences of irrigation and fertilization on soybean properties and nitrogen leaching. *Cer. Res. Commun.* **2006**, *34*, 513–516. [CrossRef]
12. Nemcic, J.; Mesić, M.; Bašić, F.; Kisić, I.; Zgorelec, Z. Nitrate concentration in drinking water from wells at three different locations in Northwest Croatia. *Cer. Res. Commun.* **2007**, *35*, 533–536. [CrossRef]
13. Hanson, B.; Schwankl, L.; Granttan, S.; Prichard, T. *Drip Irrigation for Row Crops*; Water Management Handbook Series (Publication 93-05); University of California: Davis, CA, USA, 1996.
14. Simunek, J.; van Genuchten, M.T.; Šejna, M. Recent Developments and Applications of the HYDRUS Computer Software Packages. *Vadose Zone J.* **2016**. [CrossRef]
15. Cote, C.M.; Bristow, K.L.; Charlesworth, P.B.; Cook, F.J.; Thorburn, P.J. Analysis of soil wetting and solute transport in subsurface trickle irrigation. *Irrig. Sci.* **2003**, *22*, 143–156. [CrossRef]
16. Gardenas, A.I.; Hopman, J.W.; Hanson, B.R.; Simunek, J. Two-dimensional modeling of nitrate leaching for various fertigation scenarios under micro-irrigation. *Agric. Water Manag.* **2005**, *74*, 219–242. [CrossRef]
17. Ajdary, K.; Singh, D.K.; Singh, A.K.; Khanna, M. Modeling of nitrogen leaching from experimental onion field under drip fertigation. *Agric. Water Manag.* **2007**, *89*, 15–28. [CrossRef]
18. Crevoisier, D.; Popova, Z.; Mailhol, J.C.; Ruelle, P. Assessment and simulation of water and nitrogen transfer under furrow irrigation. *Agric. Water Manag.* **2008**, *95*, 354–366. [CrossRef]
19. Siyal, A.A.; Bristow, K.L.; Simunek, J. Minimizing nitrogen leaching from furrow irrigation through novel fertilizer placement and soil surface management strategies. *Agric. Water Manag.* **2012**, *115*, 242–251. [CrossRef]
20. Ramos, T.B.; Simunek, J.; Goncalves, M.C.; Martins, J.C.; Prazeres, A.; Pereira, L.S. Two-dimensional modeling of water and nitrogen fate from sweet sorghum irrigated with fresh and blended saline waters. *Agric. Water Manag.* **2012**, *111*, 87–104. [CrossRef]
21. Siyal, A.A.; Siyal, A.G. Strategies to reduce nitrate leaching under furrow irrigation. *Int. J. Environ. Sci. Dev.* **2013**, *4*, 431–434. [CrossRef]
22. Zhen, W.; Jiusheng, L.; Yanfeng, L. Simulation of nitrate leaching under varying drip system uniformities and precipitation patterns during the growing season of maize in the North China Plain. *Agric. Water Manag.* **2014**, *142*, 19–28.
23. Raij, I.; Simunek, J.; Ben-Gal, A.; Lazarovitch, N. Water flow and multicomponent solute transport in drip irrigated lysimeters: Experiments and modeling. *Water Resour. Res.* **2016**, *52*, 6557–6574. [CrossRef]
24. Salehi, A.A.; Navabian, M.; Varaki, M.E.; Pirmoradian, N. Evaluation of HYDRUS-2D model to simulate the loss of nitrate in subsurface controlled drainage in a physical model scale of paddy fields. *Padd. Water Environ.* **2017**, *15*, 433–442. [CrossRef]

25. Shekofteh, H.; Afyuni, M.; Hajabbasi, M.A. Modeling of nitrate leaching from a potato field using HYDRUS-2D. *Commun. Soil Sci. Plant Anal.* **2013**, *44*, 2917–2931. [CrossRef]

26. Mokhtar, S.; El Agroudy, N.; Shafiq, F.A.; Abdel Fatah, H.Y. The Effects of the environmental pollution in Egypt. *Intern. J. Environ.* **2015**, *4*, 21–26.

27. Hillel, D. *Environmental Soil Physics*; Elsevier: San Diego, CA, USA, 1998.

28. Abou Lila, T.S.; Berndtsson, R.; Persson, M.; Somaida, M.; El_Kiki, M.; Hamed, Y.; Mirdan, A. Numerical evaluation of subsurface trickle irrigation with brackish water. *Irrig Sci.* **2013**, *31*, 1125–1137. [CrossRef]

29. Anderson, M.P. Movement of contaminants in groundwater: Groundwater transport-advection and dispersion. In *Groundwater Contamination. Studies in Geophysics*; National Academy: Washington, DC, USA, 1984; pp. 37–45.

30. Cote, C.M.; Bristow, K.L.; Ford, E.J.; Verburg, K.; Keating, B. *Measurement of Water and Solute Movement in Large Undisturbed Soil Cores: Analysis of Macknade and Bundaberg Data*; Technical Report 07/2001; CSIRO Land and Water: Canberra, Australia, 2001.

31. Lide, D.R. *CRC Handbook of Chemistry and Physics*; CRC Press: Boca Raton, FL, USA, 2005.

32. Lotse, E.G.; Jabro, J.D.; Simmons, K.E.; Baker, D.E. Simulation of nitrogen dynamics and leaching from arable soils. *J. Contam. Hydrol.* **1992**, *10*, 183–196. [CrossRef]

33. Jansson, P.E.; Karlberg, L. *Coupled Heat and Mass Transfer Model for Soil–Plant–Atmosphere Systems*; Royal Institute of Technology, Department of Civil and Environmental Engineering: Stockholm, Sweden, 2001; p. 321.

34. Flury, M.; Fluhler, H. Tracer characteristics of Brilliant Blue FCF. *Soil Sci. Soc. Am. J.* **1995**, *59*, 22–27. [CrossRef]

35. Selim, T.; Berndtsson, R.; Persson, M.; Somaida, M.; El-Kiki, M.; Hamed, Y.; Mirdan, A.; Zhou, Q. Influence of geometric design of alternate partial root-zone subsurface drip irrigation (APRSDI) with brackish water on soil moisture and salinity distribution. *Agric. Water Manag.* **2012**, *103*, 182–190. [CrossRef]

36. Heuvelink, E. Growth, development and yield of a tomato crop: periodic destructive measurements in a greenhouse. *Hort. Sci.* **1995**, *61*, 77–99. [CrossRef]

37. Vrugt, J.A.; van Wijk, M.T.; Hopmans, J.W.; Simunek, J. Comparison of one, two, and three-dimensional root water uptake functions for transient water flow modeling. *Water Resour. Res.* **2001**, *37*, 2457–2470. [CrossRef]

38. Feddes, R.A.; Kowalik, P.J.; Zaradny, H. *Simulation of Field Water Use and Crop Yield*; John Wiley & Sons: New York, NY, USA, 1978.

© 2019 by the authors. Licensee MDPI, Basel, Switzerland. This article is an open access article distributed under the terms and conditions of the Creative Commons Attribution (CC BY) license (http://creativecommons.org/licenses/by/4.0/).

Article

Comparison of Soil EC Values from Methods Based on 1:1 and 1:5 Soil to Water Ratios and EC_e from Saturated Paste Extract Based Method

George Kargas, Paraskevi Londra * and **Anastasia Sgoubopoulou**

Laboratory of Agricultural Hydraulics, Department of Natural Resources Management and Agricultural Engineering, Agricultural University of Athens, 75 Iera Odos, 11855 Athens, Greece; kargas@aua.gr (G.K.); anasgo@aua.gr (A.S.)
* Correspondence: v.londra@aua.gr; Tel.: +30-210-529-4069

Received: 7 March 2020; Accepted: 31 March 2020; Published: 2 April 2020

Abstract: The present study investigates the effect of three different methods of obtaining 1:1 and 1:5 soil-over-water mass ratios (soil:water) extracts for soil electrical conductivity (EC) measurements ($EC_{1:1}$, $EC_{1:5}$). On the same soil samples, also the electrical conductivity of the saturated paste extract (EC_e) was determined and the relationships between EC_e and each of the three of $EC_{1:1}$ and $EC_{1:5}$ values were examined. The soil samples used were collected from three areas over Greece (Laconia, Argolida and Kos) and had EC_e values ranging from 0.611 to 25.9 dS m^{-1}. From the results, it was shown that for soils with $EC_e < 3$ dS m^{-1} the higher EC values were obtained by the method where the suspension remained at rest for 23 hours and then shaken mechanically for 1 h. On the contrary, no differences were observed among the three methods for soils with $EC_e > 3$ dS m^{-1}. Also, in the case of $EC_{1:5}$, the optimal times for equilibration were much longer when $EC_e < 3$ dS m^{-1}. Across all soils, the relationships between EC_e and each of three methods of obtaining $EC_{1:1}$ and $EC_{1:5}$ were strongly linear ($0.953 < R^2 < 0.991$ and $0.63 < $ RMSE $ < 1.27$ dS m^{-1}). Taking into account the threshold of $EC_e = 3$ dS m^{-1}, different $EC_e = f(EC_{1:5})$ linear relationships were obtained. Although the linear model gave high values of R^2 and RMSE for $EC_e < 3$ dS m^{-1}, the quadratic model resulted in better R^2 and RMSE values for all methods examined. Correspondingly, in the 1:1 method, two of the three methods used exhibited similar slope values of the linear relationships independent of EC_e value ($EC_e < 3$ or $EC_e > 3$ dS m^{-1}), while one method (23 h rest and then shaken mechanically for 1 hour) showed significant differences in the slopes of the linear relationships between the two ranges of EC_e.

Keywords: saturated soil paste; electrical conductivity; salinity

---

1. Introduction

Soil salinity is one of the basic limiting factors in food production especially in arid and semi-arid regions since most crops are sensitive to increased salt concentration in the soil solution [1]. Soil salinization is particularly acute in arid and semi-arid areas with shallow groundwater as well irrigation water of poor quality.

Soil salinity assessment is based on measurement of the electrical conductivity of soil saturated paste extract (EC_e); this has been established as the standard method [2,3]. Saline soils are considered to be the soils where the saturated paste extract has EC_e values greater than 4 dS m^{-1}. However, this method is laborious and time consuming especially in the case of EC_e determination for a large number of soil samples. Additionally, the method appears to be more difficult and requires skills and expertise to obtain saturation point for clay soils.

For these reasons, many researchers have suggested easier methods to determine EC in various soils over water mass ratios extracts instead of determining EC_e. The most widely used soil over water mass ratios, (soil:water), are the 1:1 and the 1:5. The ratio of 1:5 is used for soil salinity assessment ($EC_{1:5}$) in Australia and China [4,5], while the ratio 1:1 ($EC_{1:1}$) is commonly used in the United States [6]. Therefore, different methods for EC assessment are applied between different regions and organizations.

Many researchers have proposed linear relationships between EC_e and $EC_{1:1}$ or $EC_{1:5}$ [7], (Table 1). However, the coefficients of the linear relationships are different and vary according to the area of interest. These coefficients are affected, among other factors, by the soil texture [8–10], the presence of gypsum and calcite in the soil [3,11], the chemical composition of the soil solution, the cation exchange capacity, etc. It has been documented that in the case of coarse-textured soils the slopes of the abovementioned linear relationships is greater than those of fine-textured soils [8].

The equilibration time and the method of preparation and extraction for determining $EC_{1:1}$ or $EC_{1:5}$ are probably additional factors that have led to the observed differences among various models [6,12]. It is worth to know that the equations $EC_e = f(EC_{1:5})$ and $EC_e = f(EC_{1:1})$ presented in Table 1 are often compared without taking into account these factors even though the equations have been obtained by different methods and at different ranges of EC_e values. More specific, Aboukila and Norton [13] and Aboukila and Abdelaty [14] have used the NRCS method [15], Khorsandi and Yazdi [11] have shaken the suspension for 1 h, Sonmez et al. [10] have used the USDA method [16], while Visconti et al. [3] have applied mechanical shake for 24 h (Table 1). As regards to the EC_e values range, Aboukila and Norton [13] presented their equation for EC_e values up to 10.26 dS m^{-1}, while Zhang et al. [17] and Khorsandi and Yazdi [11] for EC_e values up to 108 and 170 dS m^{-1}, respectively (Table 1). Noted that such extreme EC_e values are related to very specific cases (e.g., dumping of saline water as waste from the oil industry or saline areas for large scale halophyte production). Overall, to obtain the equations $EC_e = f(EC_{1:5})$ and $EC_e = f(EC_{1:1})$ both different methods have been applied to measure $EC_{1:5}$ and $EC_{1:1}$ and different ranges of EC_e values.

He et al. [6] reported that the $EC_{1:5}$ was affected by both agitation method and agitation time. Specifically, significant differences existed within three agitation methods when EC_e values ranged between 0.96 and 21.2 dS m^{-1}. Equilibration times were significantly greater for soils having $EC_e < 4$ dS m^{-1} compared to soils having $EC_e > 4$ dS m^{-1}. The agitation method of shaking plus centrifuging showed the greatest values of $EC_{1:5}$ while the stirring method showed the smallest ones for the same soil examined. Also, Vanderheynst et al. [12], conducting an experiment with compost using various dilutions, found that as agitation time increased the EC values increased—especially when agitation time increased from 3 to 15 h. The above results showed the important role of agitation time among the different agitation methods on EC measurement, irrespective of the porous medium (e.g., soil, compost).

Among the various methods widely used—especially in the case of 1:5 ratio—there are the following three methods:

(i)   Loveday [18]: the suspension is mechanically shaken for 1 h and then kept at rest for 20 min.
(ii)   NRCS [15]: the suspension remains at rest in complete shade for 23 h and then shaken mechanically for 1 h.
(iii) USDA [2]: the suspension is shaken by hand, 4 times, every 0.5 h for 30 s.

The difference between methods (i) and (ii) lies in the different rest times of the suspension, while methods (i) and (ii) differ from (iii) in both the shaking mode and the rest time.

Table 1. Relationships between soil saturated paste extract electrical conductivity (EC_e) and 1:1 and 1:5 soil to water extract electrical conductivities ($EC_{1:1}$, $EC_{1:5}$) as proposed by several researchers, as well as the extraction method and the corresponding range of EC_e values.

Reference	Expression	Method	EC_e Values Range (dS m^{-1})
USDA [16]	$EC_e = 3\,(EC_{1:1})$ [f]		
Khorsandi and Yazdi [11]	$EC_e = 7.94\,(EC_{1:5}) + 0.27$ [d] $EC_e = 9.14\,(EC_{1:5}) - 15.72$ [e]	Shake 1 h	1.04–170
Sonmez et al. [10]	$EC_e = 2.03\,(EC_{1:1}) - 0.41$ [c] $EC_e = 7.36\,(EC_{1:5}) - 0.24$ [c]	Rhoades [19]	0.22–17.68
Frazen [9]	$EC_e = 2.96\,(EC_{1:1}) - 0.95$ [c]	N/A	N/A
Aboukila and Norton [13]	$EC_e = 5.04\,(EC_{1:5}) + 0.37$ [c]	NRCS method [15]	0.624–10.26
Chi and Wang [20]	$EC_e = 11.74\,(EC_{1:5}) - 6.15$ [b] $EC_e = 11.04\,(EC_{1:5}) - 2.41$ [c] $EC_e = 11.68\,(EC_{1:5}) - 5.77$ [f]	USDA method [16]	1.02–227
Slavich and Petterson [8]	$EC_e = f(EC_{1:5})$	Loveday [18]	0–38
Ozcan et al. [21]	$EC_e = 1.93\,(EC_{1:1}) - 0.57$ [f] $EC_e = 5.97\,(EC_{1:5}) - 1.17$ [f]	N/A	N/A
Aboukila and Abdelaty [14]	$EC_e = 7.46\,(EC_{1:5}) + 0.43$ [a]	NRCS method [15]	0–18.3
Hong and Henry [22]	$EC_e = 1.56\,(EC_{1:1}) - 0.06$ [f]	Shake 1 h	0.25–42.01
Zhang et al. [17]	$EC_e = 1.79\,(EC_{1:1}) + 1.46$ [f]	Equilibrate 4 h	0.165–108
Visconti et al. [3]	$EC_e = 5.7\,(EC_{1:5}) - 0.2$	Shake 24 h	0.5–14
Kargas et al. [7]	$EC_e = 1.83\,(EC_{1:1}) + 0.117$ [c] $EC_e = 6.53\,(EC_{1:5}) - 0.108$ [c]	USDA [16]	0.47–37.5

The indices a, b and c refer to coarse, medium and fine soils, respectively. The indices d and e refer to the presence or absence of gypsum, respectively. The index f refers to combined soil texture.

Still now, no comparison has been made among the three abovementioned widely spread EC methods. Also, from international literature, it seems that there is no research work referred on the effect of different methods on the $EC_{1:1}$, although different methods have been used on the $EC_{1:1}$ measurement [16,17].

The objectives of present work are: (i) The comparison of EC values derived from the three most commonly used methods of 1:1 and 1:5 extracts; to investigate whether the differences between these methods are maintained across a range of soil EC_e and (ii) the investigation of the relationship between EC_e and EC values derived from the three methods.

2. Materials and Methods

2.1. Sample Collection Areas

The soil samples examined were collected from three areas in Greece, and more specifically, from the Prefectures of Lakonia, Argolida and from the island of Kos. Specifically, 50 soil samples were collected from Laconia from irrigated olive groves. The sampling procedure was carried out in September after the irrigation period. In Argolida, 12 samples were collected from various irrigated crops at the end of the irrigation period, while in Kos, 27 samples were collected from a horticultural greenhouse. The depth of soil samples collection was up to 30 cm.

2.2. Methods of Determining the Soil Properties

After sampling, the samples were transferred to the laboratory for air-drying and sieving through a 2 mm sieve and the soil texture, pH and calcium carbonate were determined. Soil texture was determined by means of the Bouyoucos hydrometer method [23], pH values were measured using standard glass/calomel electrodes in 1:2.5 w/v soil–water suspension [24]; $CaCO_3$ equivalent percentage was estimated by measuring the eluted CO_2 following the addition of HCl (calcimeter Bernard method).

2.3. Methods of Various Soil Extraction and Measurements

2.3.1. EC_e Method

350 g of soil was used to prepare the soil saturated paste and then the paste was allowed to stand for 24 h (USDA, 1954). Subsequently, the vacuum extracts were collected and EC_e was measured by a conductivity meter (WTW, Cond 315i). For the saturation percentage (SP) determination, a subsample of each paste was oven dried at 105 °C for 24 h.

2.3.2. $EC_{1:5}$ Method

For the 1:5 suspension, 50 g of soil and 250 mL of distilled water were used. Three alternative methods were applied: the method of Loveday [18], the NRCS [15] and the USDA [2].

In the Loveday method, the suspension was shaken by a mechanical shaker for exactly one hour and then kept at rest for 20 min. After the rest time, the extract was obtained, and the EC was determined. For the NRCS method, the suspension remains at rest in complete shade for 23 h and then shaken mechanically for one hour. After the shaking, the extract was obtained, and the EC was determined. Finally, in the USDA method the suspension was shaken by hand, 4 times, every half hour for 30 s. After, the extract was obtained, and the EC was determined. The method of vacuum filtration in all the three methods is the same and common, followed by the measurement of EC with a conductivity meter. All the methods and EC readings were conducted at 25 °C.

In two soil samples, one from Laconia (sample L) and one from Argolida (sample A) with EC_e values of 0.793 and 13.78 dS m^{-1}, respectively, the $EC_{1:5}$ values were measured after the suspensions were agitated with mechanical shaker for times 1, 2, 3, 4, 6, 24 and 48 h. After each agitation time the extraction was obtained, and the EC was determined. This process can better evaluate the role of shaking time on the $EC_{1:5}$ values for the two very different EC_e values.

2.3.3. $EC_{1:1}$ Method

In the 1:1 method, the three above mentioned methods (Loveday, NRCS and USDA) were also applied as described in the 1:5 method. For each of the above methods, 50 g of soil was weighed and then each procedure was performed in the same way as above.

2.3.4. Statistical Analysis

For the relationships $EC_e = f(EC_{1:1})$ and $EC_e = f(EC_{1:5})$, a least-squared linear regression was applied and the coefficient of determination R^2 was evaluated. The R^2 coefficient is used to assessing the correlation between two independent methods. Also, the values of root mean square errors (RMSE) were determined. Analysis of variance (ANOVA) was applied to test the significant difference among the applied $EC_{1:5}$ or $EC_{1:1}$ methods using SPSS Statistical Software v. 17.0 (SPSS Inc., Chicago, IL, USA); the means of each method were compared using t-test at a probability level $P = 0.05$.

3. Results and Discussion

3.1. Soil Properties

Samples from Laconia and Argolida are characterized as clay-clay loam soils and from Kos as sandy clay soils. All soil samples presented negligible gypsum content. As regards to $CaCO_3$, samples from Laconia presented a content lower than 2.5%, from Argolida 5–8% and from Kos 8.5–11%. The pH values ranged from 7.69 to 8.06 for soil samples from Laconia and from 7.5 to 7.7 for soil samples from Argolida and Kos.

Additionally, the soil texture analyses of the two soil samples examined separately resulted as follows: (i) soil sample L—clay soil (23.5% sand, 16% silt, 60.5% clay) and (ii) soil sample A—clay loam/loam soil (39% sand, 32% silt, 29% clay). The $CaCO_3$ content was 0.2% and 7.66% and pH values were 7.75 and 7 for sample L and A, respectively.

3.2. Estimation of Soil Salinity

The EC_e values ranged from 0.611 to 25.9 dS m^{-1}. It should also be noted that the EC_e variation range of the soil samples from Laconia is much lower than that of the other two regions (Argolida and Kos). Specifically, EC_e values of the samples from Laconia ranged from 0.611 to 1.664 dS m^{-1}, while in the other two regions they ranged from 2.32 to 25.9 dS m^{-1}. From the measured EC_e values, it appears that a relatively wide range in salinity levels was obtained for both comparing the different $EC_{1:5}$ and $EC_{1:1}$ methods, as well as evaluating the relationship between the EC_e and each of $EC_{1:5}$ or $EC_{1:1}$ methods.

As regards to SP all soil samples examined (with exception of the two separated samples) have values greater than 43%, percentage which indicates that the soils are classified in fine textured soils [20]. More specifically, SP values ranged from 50.5% to 72.5% for soils from Laconia, 52–70% for soils from Argolida and 43–53% for soils from Kos.

3.3. Comparison of 1:1 and 1:5 Soil to Water Extract Electrical Conductivity Methods

In Table 2 the slope of the linear relationship (y = ax) between 1:5 soil to water extract electrical conductivity methods for $EC_e < 3$ dS m^{-1} and $EC_e > 3$ dS m^{-1} and R^2 are presented.

Table 2. Slopes of the linear equations describing the relation between 1:5 soil to water extract electrical conductivity methods for $EC_e < 3$ dS m^{-1} and $EC_e > 3$ dS m^{-1} and coefficient of determination R^2.

$EC_{1:5}$		
Methods	Slope	R^2
$EC_e < 3$ dS m^{-1}		
NRCS–Loveday method	1.166	0.872
NRCS–USDA	1.047	0.797
USDA–Loveday method	1.108	0.812
$EC_e > 3$ dS m^{-1}		
NRCS–Loveday method	1.01	0.990
NRCS–USDA	1.00	0.960
USDA–Loveday method	1.00	0.976

Similarly, the slope and R^2 of the linear relationship between 1:1 soil to water extract electrical conductivity methods for $EC_e < 3$ dS m^{-1} and $EC_e > 3$ dS m^{-1} are presented in Table 3.

Table 3. Slopes of the linear equations describing the relation between 1:1 soil to water extract electrical conductivity methods for $EC_e < 3$ dS m^{-1} and $EC_e > 3$ dS m^{-1} and coefficient of determination R^2.

$EC_{1:1}$		
Methods	Slope	R^2
$EC_e < 3$ dS m^{-1}		
NRCS–Loveday method	1.185	0.800
NRCS–USDA	1.161	0.781
USDA–Loveday method	1.012	0.817
$EC_e > 3$ dS m^{-1}		
NRCS–Loveday method	1.01	0.984
NRCS–USDA	0.97	0.945
USDA–Loveday method	1.09	0.952

From the results presented in Tables 2 and 3, it is obvious that each of the three methods examined resulted in different values of both $EC_{1:1}$ and $EC_{1:5}$ when $EC_e < 3$ dS m^{-1}. Analysis of variance (ANOVA) showed that the three methods are significantly different at a probability level P = 0.05. Furthermore, the t-test analysis (P = 0.05) showed that the NRCS and Loveday methods as well as the USDA and Loveday methods resulted in significantly different $EC_{1:5}$ values, while $EC_{1:5}$ values between NRCS and USDA were not significantly different. The mean value with standard deviation for NRCS, USDA and Loveday methods were 0.177 ± 0.029, 0.169 ± 0.029 and 0.151 ± 0.027 dS m^{-1}, respectively. In the case of 1:1 ratio, the EC values between NRCS and USDA as well as NRCS and Loveday methods were also significantly different (P = 0.05). The mean value with standard deviation for NRCS, USDA and Loveday methods were 0.5 ± 0.070, 0.43 ± 0.100 and 0.423 ± 0.086 dS m^{-1}, respectively.

The NRCS method resulted in greater EC values compared to the other two methods for both 1:1 and 1:5 ratios, whereas the Loveday method resulted in lower EC values. From these results, it appears that at low values of EC_e ($EC_e < 3$ dS m^{-1}) the rest time seems to play an important role since the difference between the NRCS and the Loveday method is only in the duration of rest time. As regards to the NRCS and USDA methods, the slope of the linear regression between the NRCS and USDA at 1:5 ratio is 1.047, while at 1:1 is 1.161.

The $EC_{1:5}$ values of the soil sample L (with $EC_e = 0.793$ dS m^{-1} < 3 dS m^{-1}) obtained by mechanical shaking for 1, 2, 3, 4 and 6 h was approximately 0.142 dS m^{-1} while $EC_{1:5}$ values for 24 and 48 h were 0.218 and 0.274 dS m^{-1}, respectively. Practically, after 48 h shaking the $EC_{1:5}$ value was approximately doubling. The corresponding EC values obtained by the three methods used were 0.141, 0.127 and 0.158 dS m^{-1} for USDA, Loveday and NRCS methods, respectively. Therefore, it appears that the agitation time plays a dominant role to obtain equilibrium since the difference between the NRCS method ($EC_{1:5} = 0.158$ dS m^{-1}) and the method with 24 h shaking ($EC_{1:5} = 0.218$ dS m^{-1}) is in the shaking time. These results are similar to those of He et al. [6] in terms of the long shaking time required to equilibration but differ in the fact that in our experiments did not show differences in EC values obtained by shaking of at least up to 6 h. He et al. [6] explained that the higher values of EC obtained by the long shaking time method compared to other methods may be due to the fact that the mechanical shaking destroys micro-aggregates, as well as increase dissolution of salts because the dynamic concentration gradient between solid and liquid phases. Also, Vanderheynst et al. [12] found that differences occur for shaking time greater than a threshold value of 3 h.

In the case of soils with $EC_e > 3$ dS m^{-1} there is no significant differences between agitation methods since all methods gave almost the same results and the slope of the linear relationship is almost 1 (Tables 2 and 3). In addition, it is noted that the R^2 values for soils with $EC_e > 3$ dS m^{-1} are higher for all methods examined, in both 1:5 and 1:1 ratios, compared to R^2 values for $EC_e < 3$ dS m^{-1} (Tables 2 and 3).

The $EC_{1:5}$ values of the soil sample A (with $EC_e = 13.8$ dS m^{-1} > 3 dS m^{-1}) obtained by mechanical shaking for 1, 2, 3, 4, 6, 24 and 48 h ranged from 1.683 to 1.751 dS m^{-1}. It is obvious that for soils with $EC_e > 3$ dS m^{-1} the shaking times required to obtain equilibration are significantly lower compared to soils with $EC_e < 3$ dS m^{-1}

The different behavior depending on the EC_e value shows that the solid and liquid phases is far from considered a simple system where the only process carried out is dissolution and that the concentration of ions is inversely proportional to dilution. Such situations may exist only in sandy or sandy loam soils in semi-arid areas with high salinity [25]. However, the soils are characterized by a cation exchange capacity value depending on the type and quantity of clay, the presence of slightly soluble minerals but also ion exchanges between solid and liquid phase. In the present experimental work, the existence of a relatively high clay percentage combined with the existence of slightly soluble minerals may be led to different EC values among various methods, especially when $EC_e < 3$ dS m^{-1}. This phenomenon may be even more pronounced in the case of clay soils where there are high content of slightly soluble minerals but less pronounced in the coarse-textured soils without slightly soluble minerals.

3.4. Relationship between EC_e and 1:5 Soil to Water Extract Electrical Conductivity Methods

In Table 4, the linear relationships between EC_e and $EC_{1:5}$, for all soil samples, determined by the three different methods are presented. Analysis of the results showed that each 1:5 soil to water extract electrical conductivity method is strongly related with EC_e since R^2 values are high ($0.953 < R^2 < 0.972$) and RMSE are low (1.02 dS m^{-1} < RMSE < 1.27 dS m^{-1}). It also appears that the linear equations showed small differences regardless of the $EC_{1:5}$ methods for all soils examined. These data confirm the existence of a strong linear relationship when the range of EC_e is relatively great (Table 1).

Table 4. Regression equations describing the relation between saturated paste extracts EC_e and $EC_{1:5}$ determined by three different methods with the coefficients of determination (R^2) and root mean square errors (RMSE) for all soil samples examined.

	$EC_{1:5}$		
Methods	$EC_e = fEC_{1:5}$	R^2	**RMSE (dS m^{-1})**
EC_e–NRCS	$EC_e = 6.58\ EC_{1:5}$	0.973	1.09
EC_e–USDA	$EC_e = 6.61\ EC_{1:5}$	0.953	1.27
EC_e–Loveday method	$EC_e = 6.71\ EC_{1:5}$	0.971	1.02

As shown in Table 4, the relationship $EC_e = fEC_{1:5}$ using the USDA method is similar to the corresponding one reported by Kargas et al. [7], (Table 1) for Greek soils since both the two equations have almost the same slope (6.61 and 6.53, respectively).

However, analysis of the results for soils with $EC_e < 3$ dS m^{-1} showed that a percentage of 70% of experimental EC_e values were lower than those calculated by the equations presented in Table 4. For this reason, the data were separated into two ranges based on the threshold value $EC_e = 3$ dS m^{-1} to evaluate whether the relationship $EC_e = fEC_{1:5}$ is described by different equations as reported by other researchers [26,27].

The slopes of linear equation describing the relation between EC_e and $EC_{1:5}$ determined by three different methods, as well as the R^2 and RMSE for all soil examined for $EC_e < 3$ dS m^{-1} and $EC_e > 3$ dS m^{-1}, are presented in Table 5.

Table 5. Regression equations describing the relation between saturated paste extracts EC_e and $EC_{1:5}$ determined by three different methods with the coefficients of determination (R^2) and root mean square errors (RMSE) for all soil examined for $EC_e < 3$ dS m^{-1} and $EC_e > 3$ dS m^{-1}.

	$EC_{1:5}$		
Methods	$EC_e = fEC_{1:5}$	R^2	**RMSE (dS m^{-1})**
	$EC_e < 3$ **dS m^{-1}**		
EC_e–NRCS	$EC_e = 4.68\ EC_{1:5}$	0.718	0.189
EC_e–USDA	$EC_e = 4.89\ EC_{1:5}$	0.537	0.130
EC_e–Loveday method	$EC_e = 5.46\ EC_{1:5}$	0.647	0.123
	$EC_e > 3$ **dS m^{-1}**		
EC_e–NRCS	$EC_e = 6.60\ EC_{1:5}$	0.934	1.710
EC_e–USDA	$EC_e = 6.60\ EC_{1:5}$	0.917	1.800
EC_e–Loveday method	$EC_e = 6.71\ EC_{1:5}$	0.942	1.580

As shown in Table 5, for soils with $EC_e < 3$ dS m^{-1}, the slope of the linear equation between EC_e and $EC_{1:5}$ has different value depending on $EC_{1:5}$ determination method used with the smallest and the highest values obtained by the NRCS and Loveday method. Also, the values of the slopes of linear relationships, for both $EC_e < 3$ dS m^{-1} and $EC_e > 3$ dS m^{-1}, differ significantly from each other since in the case of $EC_e < 3$ dS m^{-1} these values ranged from 4.68 to 5.46, while they ranged from 6.60 to 6.71 in the case of $EC_e > 3$ dS m^{-1}. In addition, for $EC_e < 3$ dS m^{-1} R^2 values are lower ($0.537 < R^2 < 0.718$)

than those ones ($0.917 < R^2 < 0.942$) observed for $EC_e > 3$ dS m^{-1} indicating a strong linear relation between EC_e and each $EC_{1:5}$ determination method.

Comparison between the same methods for both $EC_e < 3$ dS m^{-1} and $EC_e > 3$ dS m^{-1} showed a difference between slopes ranging from 18.5% to 28.9%. Thus, in order to compare various equations describing the relationship between EC_e and $EC_{1:5}$, both the agitation method of $EC_{1:5}$ determination and the range of EC_e for which the equation has been proposed should be taken into account. Specifically, as shown in Table 5 and Figure 1, the relationship between EC_e and $EC_{1:5}$ determined by the NRCS method has a slope of 4.68 for $EC_e < 3$ dS m^{-1} and 6.60 for $EC_e > 3$ dS m^{-1}. The differences among the methods may be even greater if the soil contains gypsum or larger amounts of calcite than those observed in the soil samples examined.

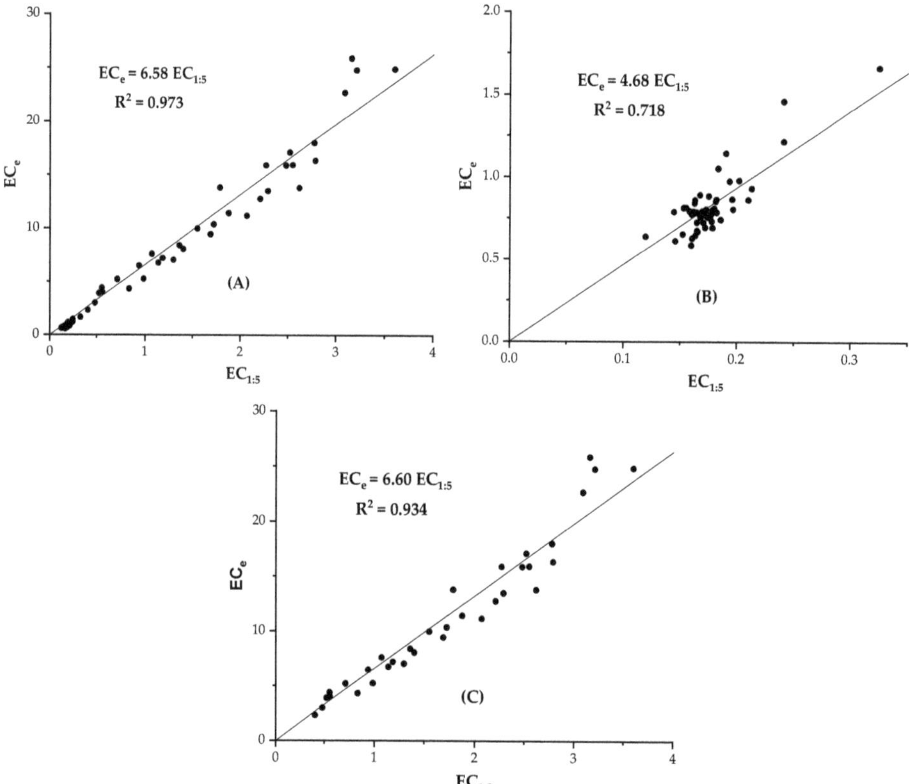

Figure 1. Relationship between EC_e and $EC_{1:5}$ for NRCS extraction method. A: all soil samples, B: soil samples range $EC_e < 3$ dS m^{-1}, C: soil samples range $EC_e > 3$ dS m^{-1}.

Similar results regarding to the effect of agitation method, the range of EC_e and the gypsum content on equation describing the relationship between EC_e and $EC_{1:5}$ have been presented by other researchers [3,26,27].

He et al. [27] proposed a quadratic equation as a more appropriate equation to describe the relationship between EC_e and $EC_{1:5}$ when EC_e values are lower than 4 dS m^{-1}. The fitting of a quadratic equation to the data of this study for $EC_e < 3$ dS m^{-1} gave R^2 values of 0.74, 0.57 and 0.66 and RMSE values 0.096 (NRCS), 0.124 (USDA) and 0.115 dS m^{-1} (Loveday method), respectively. A comparison between these RMSE values and those of the linear relationships presented in Table 5, showed a significant improvement only in the case of the NRCS method. It should be noted that there is a significant difference in RMSE values presented in Table 4 compared to RMSE values whether we use the linear equation or quadratic equation to EC_e estimation for $EC_e < 3$ dS m^{-1}.

3.5. Relationship between EC_e and 1:1 Soil to Water Extract Electrical Conductivity Methods

Table 6 shows the relationship between EC_e and the three methods of determining $EC_{1:1}$ for all soil samples examined. The results showed that the relationship is strongly linear in all methods examined ($R^2 > 0.986$) and RMSE values are low ($0.63 < $ RMSE $ < 0.74$ dS m^{-1}). The values of both R^2 and RMSE indicate that this linear relationship reliably estimates the EC_e. However, $EC_e = fEC_{1:1}$ linear relationships have different f coefficient for each method.

Table 6. Regression equations describing the relation between saturated paste extracts EC_e and $EC_{1:1}$ determined by three different methods with the coefficients of determination (R^2) and root mean square errors (RMSE) for all soil examined.

	$EC_{1:1}$		
Methods	$EC_e = fEC_{1:1}$	R^2	**RMSE (dS m^{-1})**
EC_e–NRCS	$EC_e = 2.07\ EC_{1:1}$	0.986	0.63
EC_e–USDA	$EC_e = 1.93\ EC_{1:1}$	0.991	0.74
EC_e–Loveday method	$EC_e = 2.12\ EC_{1:1}$	0.988	0.68

In Table 7, regression equations describing the relation between EC_e and $EC_{1:1}$ determined by three different methods are presented taking into consideration the threshold of EC_e value 3 dS m^{-1}. The results showed that the same trends were observed for R^2 and RMSE values as in the case of the results of 1:5 ratio presented in Table 5. As regards to differences observed in the slope of linear relationships between the two areas of EC_e values, a notable difference was observed in the NRCS method since it resulted to a slope 1.65 for $EC_e < 3$ dS m^{-1} and 2.08 for $EC_e > 3$ dS m^{-1}. Furthermore, the quadratic equation for the NRCS method, for $EC_e < 3$ dS m^{-1}, resulted almost to the same RMSE values (0.099 dS m^{-1}) with those of linear equation. Therefore, for this method with $EC_e < 3$ dS m^{-1} the simple linear equation gave quite reliable results to EC_e estimation. The other two methods showed similar slope values regardless of the EC_e value. In particular, the EC_e-USDA relationship had almost the same slope value regardless of the EC_e.

Table 7. Regression equations describing the relation between saturated paste extracts EC_e and $EC_{1:1}$ determined by three different methods with the coefficients of determination (R^2) and root mean square errors (RMSE) for all soil examined for $EC_e < 3$ dS m^{-1} and $EC_e > 3$ dS m^{-1}.

	$EC_{1:1}$		
Methods	$EC_e = fEC_{1:1}$	R^2	**RMSE (dS m^{-1})**
	$EC_e < 3$ dS m^{-1}		
EC_e–NRCS	$EC_e = 1.65\ EC_{1:1}$	0.551	0.102
EC_e–USDA	$EC_e = 1.93\ EC_{1:1}$	0.566	0.254
EC_e–Loveday method	$EC_e = 1.96\ EC_{1:1}$	0.624	0.091
	$EC_e > 3$ dS m^{-1}		
EC_e–NRCS	$EC_e = 2.08\ EC_{1:1}$	0.985	1.62
EC_e–USDA	$EC_e = 1.90\ EC_{1:1}$	0.991	1.06
EC_e–Loveday method	$EC_e = 2.12\ EC_{1:1}$	0.984	1.62

The relationships between EC_e and $EC_{1:1}$ determined by the NRCS method taking into consideration the threshold of EC_e value 3 dS m^{-1} are also presented in Figure 2.

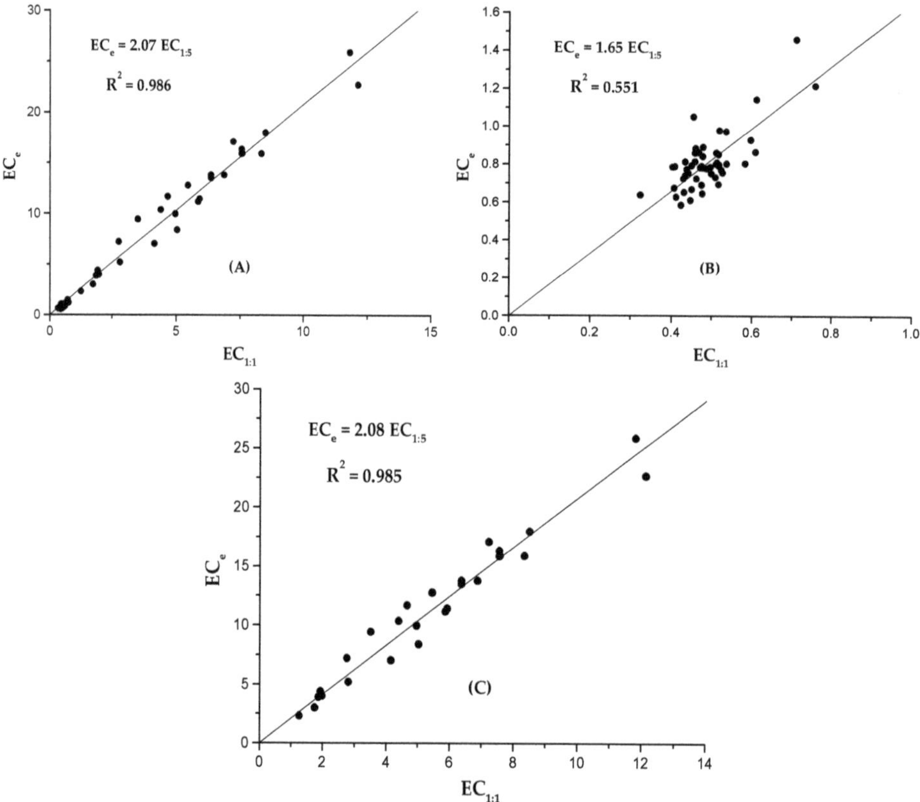

Figure 2. Relationship between EC_e and $EC_{1:1}$ for NRCS extraction method. A: all soil samples, B: soil samples range $EC_e < 3$ dS m^{-1}, C: soil samples range $EC_e > 3$ dS m^{-1}.

4. Conclusions

The $EC_{1:5}$ was affected by both agitation method and time, especially for EC_e values lower than 3 dS m^{-1}. Generally, the NRCS method resulted in the highest EC values compared to the other two methods examined. The differences among agitation methods are essentially eliminated for EC_e values greater than 3 dS m^{-1}. For soil having EC_e values lower than 3 dS m^{-1}, equilibration time was very greater than the soils having EC_e values above 3 dS m^{-1}. The most appropriate equation for EC_e estimation using $EC_{1:5}$ values for soils having $EC_e < 3$ dS m^{-1} is a quadratic equation—especially in the case of the NRCS method—while for soils having $EC_e > 3$ dS m^{-1} is the linear equation. However, if soils have a wide range of salinization levels, the linear model are recommended.

The present study shows that the shaking method and the equilibration time are additional contributing factors to the observed differences of the proposed equations for the EC_e estimation by $EC_{1:5}$. Therefore, in order to select each time, the appropriate method and equilibration time for measuring $EC_{1:5}$, during laboratory studies, the EC_e value of some samples, as well as the soil characteristics (e.g., gypsum and calcium carbonate content) should be examined in advance.

The $EC_{1:1}$ was affected by EC_e values only in the case of the NRCS method where the estimation of the EC_e can be conducted by simple but different linear relationships whose slopes depend on EC_e values. In the other two methods, the linear relationship $EC_e = f(EC_{1:1})$ was not affected by EC_e values.

Overall, it is necessary to describe in detail the method of preparation and extraction for determining $EC_{1:1}$ or $EC_{1:5}$ and the range of EC_e in order to properly evaluate and compare the proposed equations of $EC_e = f(EC_{1:5})$. Additionally, the study of soils with different characteristics than those of the group of soils examined in this work is needed.

Author Contributions: Conceptualization, G.K., P.L. and A.S.; Formal analysis, G.K., P.L. and A.S.; Methodology, G.K., P.L. and A.S.; Writing—review & editing, G.K., P.L. and A.S. All authors have read and agreed to the published version of the manuscript.

Funding: This research received no external funding.

Conflicts of Interest: The authors declare no conflict of interest.

References

1. Libutti, A.; Cammerino, A.R.B.; Monteleone, M. Risk assessment of soil salinization due to tomato cultivation in Mediterranean climate conditions. *Water* **2018**, *10*, 1503. [CrossRef]

2. USDA-Natural Resources Conservation Service. *Soil Survey Laboratory Information Manual*; Burt, R., Ed.; Soil Survey Investigations Report No. 45; version 2.0.; Aqueous Extraction, Method 4.3.3.; USDA-NRCS: Lincoln, NE, USA, 2011; p. 167.

3. Visconti, F.; De Paz, J.M.; Rubio, J.L. What information does the electrical conductivity of soil water extracts of 1 and 5 ratio (*w/v*) provide for soil salinity assessment of agricultural irrigated lands? *Geoderma* **2010**, *154*, 387–397. [CrossRef]

4. Rayment, G.E.; Lyons, D.J. *Soil Chemical Analysis Methods-Australia*; CSIRO Publishing: Collingwood, Australia, 2011.

5. Wang, Y.; Wang, Z.X.; Lian, X.J.; Xiao, H.; Wang, L.Y.; He, H.D. Measurements of soil electrical conductivity in Tianjin coastal area. *Tianjin Agric. Sci.* **2011**, *17*, 18–21.

6. He, Y.; DeSutter, T.; Prunty, L.; Hopkins, D.; Jia, X.; Wysocki, D.A. Evaluation of 1:5 soil to water extract electrical conductivity methods. *Geoderma* **2012**, *185–186*, 12–17. [CrossRef]

7. Kargas, G.; Chatzigiakoumis, I.; Kollias, A.; Spiliotis, D.; Massas, I.; Kerkides, P. Soil Salinity Assessment Using Saturated Paste and Mass Soil:Water 1:1 and 1:5 Ratios Extracts. *Water* **2018**, *10*, 1589. [CrossRef]

8. Slavich, P.G.; Petterson, G.H. Estimation the Electrical Conductivity of Saturated Paste Extracts from 1:5 Soil: Water Suspensions and Texture. *Aust. J. Soil Res.* **1993**, *31*, 73–81. [CrossRef]

9. Franzen, D. *Managing Saline Soils in North Dakota*; North Dakota State University Extension Service: Fargo, ND, USA, 2003; Available online: https://www.ag.ndsu.edu/publications/crops/managing-saline-soils-in-north-dakota (accessed on 1 April 2020).

10. Sonmez, S.; Buyuktas, D.; Okturen, F.; Citak, S. Assessment of different soil to water ratios (1:1, 1:2.5, 1:5) in soil salinity studies. *Geoderma* **2008**, *144*, 361–369. [CrossRef]

11. Khorsandi, F.; Yazdi, F.A. Gypsum and Texture Effects on the Estimation of Saturated Paste Electrical Conductivity by Two Extraction Methods. *Commun. Soil Sci. Plant Anal.* **2007**, *38*, 1105–1117. [CrossRef]

12. Vanderheynst, J.S.; Pettygrave, S.; Dooley, T.M.; Arnold, K.A. Estimating electrical conductivity of compost extracts at different extraction ratios. *Compost Sci. Util.* **2004**, *12*, 202–207. [CrossRef]

13. Aboukila, E.F.; Norton, J.B. Estimation of Saturated Soil Paste Salinity from Soil-Water Extracts. *Soil Sci.* **2017**, *182*, 107–113. [CrossRef]

14. Aboukila, E.F.; Abdelaty, E.F. Assessment of Saturated Soil Paste Salinity from 1:2.5 and 1:5 Soil-Water Extracts for Coarse Textured Soils. *Alex. Sci. Exch. J.* **2018**, *38*, 722–732. [CrossRef]

15. NRCS. Method 4.3.3: Aqueous Extraction. In *Soil Survey Laboratory Information Manual*; Burt, R., Ed.; Version 2.0.; NRCS: Washington, DC, USA, 2011.

16. United States Department of Agriculture (USDA). *Diagnoses and Improvement of Saline and Alkali Soils*; Agriculture Handbook No. 60; United States Department of Agriculture: Washington, DC, USA, 1954.

17. Zhang, H.; Schroder, J.L.; Pittman, J.J.; Wang, J.J.; Payton, M.E. Soil Salinity Using Saturated Paste and 1:1 Soil to water extract. *Soil Sci. Soc. Am. J.* **2005**, *69*, 1146–1151. [CrossRef]

18. Loveday, J. *Methods for Analysis of Irrigated Soils*; Tech. Comm. No.54, Commonwealth Bureau of Soils; Commonwealth Agricultural Bureau: Farnham Royal, UK, 1974.

19. Rhoades, J.D. Soluble Salts. In *Methods of Soil Analysis, Part 2: Chemical and Microbiological Properties*, 2nd ed.; Page, A.L., Ed.; Agronomy Monograph No. 9; American Society of Agronomy: Madison, WI, USA, 1982; pp. 167–179.

20. Chi, M.C.; Wang, Z.C. Characterizing salt affected soils of Songnen Plain using saturated paste and 1:5 soil to water extraction methods. *Arid Land Res. Manag.* **2010**, *24*, 1–11. [CrossRef]

21. Ozcan, H.; Ekinci, H.; Yigini, Y.; Yuksel, O. Comparison of four soil salinity extraction methods. In Proceedings of the 18th International Soil Meeting on Soil Sustaining Life on Earth, Managing Soil and Technology, Sanlıurfa, Turkey, 22–26 May 2006; pp. 697–703.

22. Hogg, T.J.; Henry, J.L. Comparison of 1:1 and 1:2 suspensions and extracts with the saturation extracts in estimating salinity in Saskatchewan. *Can. J. Soil Sci.* **1984**, *64*, 699–704. [CrossRef]

23. Bouyoucos, G.H. A recalibration of the hydrometer method for making mechanical analysis of soils. *Agron. J.* **1951**, *43*, 434–438. [CrossRef]

24. McLean, E.O. Soil pH and Lime Requirement. *Agron. Monogr.* **1983**, *9*, 199–223.

25. Nadler, A. Discrepancies between soil solute concentration estimates obtained by TDR and aqueous extracts. *Aust. J. Soil. Res.* **1997**, *35*, 527–537. [CrossRef]

26. Agarwal, R.R.; Das, S.K.; Mehrota, G.L. Interrelationship between electrical conductivity of 1:5 and saturation extracts and total soluble salts in saline alkali soils of the Gangetic alluvium in Uttar Pradesh. *Indian J. Agric. Sci.* **1961**, *31*, 284–294.

27. He, Y.; DeSutter, T.; Hopkins, D.; Jia, X.; Wysocki, D.A. Predicting ECe of the saturated paste extract from value of EC1:5. *Can. J. Soil Sci.* **2013**, *93*, 585–594. [CrossRef]

© 2020 by the authors. Licensee MDPI, Basel, Switzerland. This article is an open access article distributed under the terms and conditions of the Creative Commons Attribution (CC BY) license (http://creativecommons.org/licenses/by/4.0/).

 *water*

Article

The Effect of Soil Iron on the Estimation of Soil Water Content Using Dielectric Sensors

George Kargas [1], Paraskevi Londra [1,*], Marianthi Anastasatou [2] and Nick Moustakas [3]

[1]   Laboratory of Agricultural Hydraulics, Department of Natural Resources Management and Agricultural Engineering, Agricultural University of Athens, 75 Iera Odos, PC 11855 Athens, Greece; kargas@aua.gr
[2]   Laboratory of Economic Geology and Geochemistry, Department of Geology and Geoenvironment, National and Kapodistrian University of Athens, PC 15784 Athens, Greece; anastasat@geol.uoa.gr
[3]   Laboratory of Soil Science and Agricultural Chemistry, Department of Natural Resources Management and Agricultural Engineering, Agricultural University of Athens, 75 Iera Odos, PC 11855 Athens, Greece; nmoustakas@aua.gr
*   Correspondence: v.londra@aua.gr; Tel.: +30-210-529-4069

Received: 21 November 2019; Accepted: 20 February 2020; Published: 22 February 2020

Abstract: Nowadays, the estimation of volumetric soil water content (θ) through apparent dielectric permittivity (ε_a) is the most widely used method. The purpose of this study is to investigate the effect of the high iron content of two sandy loam soils on estimating their water content using two dielectric sensors. These sensors are the WET sensor operating at 20 MHz and the ML2 sensor operating at 100 MHz. Experiments on specific soil columns, in the laboratory, by mixing different amounts of water in the soils to obtain a range of θ values under constant temperature conditions were conducted. Analysis of the results showed that both sensors, based on manufacturer calibration, led to overestimation of θ. This overestimation is due to the high measured values of ε_a by both sensors used. The WET sensor, operating at a lower frequency and being strongly affected by soil characteristics, showed the greatest overestimation. The difference of ε_a values between the two sensors ranged from 14 to 19 units at the maximum actual soil water content (θ_m). Compared to the Topp equation, the WET sensor measures 2.3 to 2.8 fold higher value of ε_a. From the results, it was shown that the relationship θ_m-$\varepsilon_a^{0.5}$ remained linear even in the case of these soils with high iron content and the multi-point calibration (CALALL) is a good option where individual calibration is needed.

Keywords: apparent dielectric permittivity; soil water content; dielectric sensor; specific calibration

---

1. Introduction

The knowledge of volumetric soil water content (θ) is crucial in the estimating of soil profile water balance and in the study of transport of salts and various agrochemicals in soils, as well as in irrigation water management. In recent decades, technologies for measuring water content in porous media have been remarkably developed. This development is based on the ability to measure soil apparent dielectric permittivity, ε_a, and to the fact that there is a relationship between ε_a and actual volumetric soil water content, θ_m. Relative complex permittivity (ε_r^*) is composed of a real component and an imaginary component Equation (1):

$$\varepsilon_r^* = \varepsilon_r' - j\varepsilon_r'',\tag{1}$$

where ε_r' is the dielectric constant (the real part of relative permittivity), ε_r'' (the imaginary part of relative permittivity) is the equivalent dielectric loss taking the conductive loss into consideration, and $j = (-1)^{1/2}$. Volumetric soil water content is more directly related to ε_r'. For the Time Domain Reflectrometry (TDR) technique the real component ε_r' of the relative permittivity is considered about equal to the ε_a. In case of inorganic porous media, it has been shown that ε_a can be precisely correlated with θ using the Topp equation Equation (2) [1,2]:

$$\theta = -5.3 \times 10^{-2} + 2.92 \times 10^{-2}\varepsilon_a - 5.5 \times 10^{-4}\varepsilon_a^2 - 4.3 \times 10^{-6}\varepsilon_a^3, \tag{2}$$

otherwise

$$\varepsilon_a = 3.03 + 9.3\theta + 146\theta^2 - 76.7\theta^3 \tag{3}$$

Using TDR devices, the ε_a is measured through the transmission time of the electromagnetic wave along the device's pins inserted into the soil. Since the dielectric constant is 80 for water, 1 for air and 2–5 for solid components of the soil, it is clear that the ε_a value of soil is mainly determined by θ.

Topp and Reynolds [3] gave a linear approach of the Equation (2):

$$\theta = 0.115\varepsilon_a^{0.5} - 0.176, \tag{4}$$

In Equation (4), the slope of the linear relationship is related to the influence of clay and soil salinity, while the constant term is related to the electrical properties of the solid soil components. Soil salinity and clay percentage can affect the linearity of the relationship [4].

The high cost of TDR devices and the difficulties in waveform analysis have led to the development of commercial soil water sensors which calculate the θ through the ε_a using a different measurement way of ε_a [5–7]. Commercial dielectric sensors can be broadly classified as Frequency Domain (FD) or Amplitude Domain Reflectometry (ADR) sensors. The FD sensors include the Frequency Domain Reflectometry (FDR) and the capacitance sensors. The FDR sensor sends an electromagnetic wave into the measured medium (soil) and measures the frequency of the reflected wave, which varies with ε_a and θ [8]. With capacitance sensors, a medium's ε_a is determined by measuring the charge time of a capacitor consisting of that medium, confined by the sensor's prongs. ADR sensors infer the ε_a through measurement of voltage amplitude. The effective frequency (f) of commercial sensors, which ranges from approximately 10 to 150 MHz, is almost invariably lower than that of TDR, which ranges from approximately 300 to 1000 MHz [9]. The WET sensor (20 MHz), which belongs to the FDR category [10], calculates ε_a, soil bulk electrical conductivity and temperature in the same soil volume, while the ML2 sensor (100 MHz), which is based on ADR method, calculates only the ε_a utilizing the principle of the stationary wave [11,12]. These sensors have been used in many soil types and the relationship θ_m-$\varepsilon_a^{0.5}$ has been shown to be linear with the slope and intercept coefficients vary depending on the soil type. Only in the case of sandy soils these coefficients are practically the same among different operating frequency sensors. However, it is known that soil properties such as type and amount of clay minerals, organic matter, content and forms of iron oxides affect the value of ε_a and consequently the sensor calibration equation. The soil type and especially the increased content of clay (mainly the clay type 2:1) significantly affect the ε_a value measured using low frequency sensors. Campbell [13] investigated the dielectric behavior of various soil types in a frequency range of 1–50 MHz and reported that the lower the frequency, the greater dependence on the soil type and the higher the apparent dielectric permittivity value.

From the abovementioned soil properties, the content and forms of soil iron oxides have not been extensively studied for their effects on ε_a and consequently on the estimation of θ. Roth et al. Study [14] presented results for the relationship θ_m-ε_a obtained by TDR measurements for two soils (Rhodic ferralsol) from Brazil. These soils contained 18.3% and 18.5% total iron, respectively, with unknown percentages of maghemite and magnetite. Results showed that the relationship θ_m-ε_a has a smaller slope compared to inorganic soils up to a $\theta_m = 0.2$ m^{3}m^{-3}, while at higher soil water contents there is a convergence trend. However, these soils have a high content of clay and so it is not easy to estimate the particular effect of high iron content.

Robinson et al. [15] studied the effect of iron oxides (magnetite, hematite and goethite) on ε_a. Specifically, percentages of 5%, 10% and 15% by weight from each oxide form in an artificial porous medium composed by sand in various soil water content regimes were examined. They reported that the magnetite content compared to hematite and goethite seriously affects on ε_a in the case of TDR used, while in the case of capacitance probe (100 MHz) important role plays the iron oxide form and the length of sensor rods used. For TDR, the goethite content up to 15% did not affect the estimation of θ, whereas a hematite content of 15% led to a 6% overestimation of θ. However, in the case of 15% magnetite content, an overestimation of θ up to 60% was recorded. This effect is more pronounced with the increase of soil water content. In the case of capacitance probe at dry conditions, i.e., $\theta_m = 0$, the effect of increased iron oxide content on ε_a is low. However, at saturation, when magnetite content is 15% w/w, the ε_a readings rose to 75 and 122 when the length of sensor rods is 10 and 5 cm, respectively, while in the pure sand the value of ε_a is 27. In the case where hematite content is 15% w/w, the increase of ε_a is minor for a 10 cm sensor rods length, while the increase of ε_a is significant (44) for a 5 cm sensor rods length. Correspondingly, in the case where goethite content is 15%, the values of ε_a are 135 and 23 when the length of sensor rods is 5 and 10 cm, respectively. Overall, from the study of Robinson et al. [15], it is clearly shown that between TDR and capacitance probe there is a completely different behavior when the content of three iron oxides in soil is high.

Van Dam et al. [16] reported from TDR field studies that goethite iron-oxide precipitates significantly lower the electromagnetic wave velocity of sediments. Measured variations in magnetic permeability do not explain this decrease. The TDR measurements show that apparent dielectric permittivity of the solid material is not altered significantly by the iron-oxide material. The amount of iron oxides appears to correlate with the volumetric water content, which is the result of differences in water retention capacity between goethite and quartz. These variations in water content control apparent dielectric permittivity and explain the observed variation in electromagnetic wave velocity.

Pettinelli et al. [17] reported that ε_a increases significantly as magnetite content increases and a complex behavior occurs, in a frequency range of 500 Hz–1MHz, which depends on different factors, i.e., shape, dimension and origin of the magnetic material. In international literature, there are no studies to investigate the effect of the increased iron content of soils on the estimation of volumetric water content for relative low-frequency sensors, such as WET (20 MHz) and ML2 (100 MHz), that are used in the application of irrigation in agricultural practice.

The purpose of this study is the investigation of the iron content effect of two sandy loam soils on the determination of their apparent dielectric permittivity and the estimation of their volumetric water content using the proposed manufactured equations of two dielectric sensors for inorganic soils. The linearity of the θ_m-$\varepsilon_a^{0.5}$ relationship is also investigated when iron content of the soil is high.

2. Materials and Methods

2.1. WET and ML2 Sensors

The WET sensor [10] is a multi-parameter sensor consisting of the main body which contains the electronics and three sharpened, stainless steel rods 6.8 cm long, 3 mm in diameter and spaced 1.5 cm apart attached to it. The sensor measures the capacitance and conductance of the medium between the rods (usually soil) from changes in a generated 20 MHz signal. It directly measures the dielectric properties of the soil by generating a 20 MHz signal, which is applied to the central rod, the bulk electrical conductivity (σ_b) and soil temperature. The WET sensor, also, estimates volumetric water content and pore water electrical conductivity.

The ML2 Theta Probe [11] consists of an input/output cable, probe body and a sensing head. The sensing head has an array of four cylindrical rods, 60 mm long and 3 mm in diameter. Three outer rods form a triangle, with the fourth in the center. The outer rods are connected to instrument ground and form an electrical shield around the central rod, which transmits the signal in continuation from the probe body. The ε_a of the porous media is calculated as

$$\varepsilon_a^{0.5} = 1.07 + 6.4V - 6.4V^2 + 4.7V^3, \tag{5}$$

where V denotes the voltage in Volts with a range of 0–1 Volts. This range corresponds to an approximate water content 0 to 0.6 m^3m^{-3} and to a maximum square root of ε_a equal to 5.77, displayed on the screen of the probe.

It estimates volumetric water content by applying a 100 MHz signal via a specially designed transmission line whose impedance is changed as the impedance of the soil changes.

These sensors are used in combination to the soil water content meter HH2 [10] through which ε_a and θ values are obtained. Volumetric water content is calculated using a simple mixing formula that relates water content to the measured ε_a of the soil using the following Equation (6) [10]:

$$\theta = (\varepsilon_a^{0.5} - \alpha_0)/\alpha_1, \tag{6}$$

where α_0, α_1 are constants depended on the type of porous medium. The linearity of the θ_m-$\varepsilon_a^{0.5}$ relationship has been documented in cases where θ_m is homogeneous in the soil volume used for measurement [5,18]. For inorganic porous media, parameter values as proposed by the manufacturer are 1.8 and 10.1 for WET sensor, and 1.6 and 8.4 for ML2 sensor.

2.2. Measurement of Physical and Chemical Properties of Soils

Two sandy loam soil samples were taken from Central Greece, (Beotia), transferred to the laboratory, air dried, ground and separated into <2 mm fraction using a 2 mm (fine earth) sieve. The water content of dried samples was determined by drying at 105 °C in order to allow correction of assay results. Soil texture was estimated using the Bouyoucos hydrometer method, organic matter was determined using a modified Walkley–Black method, the $CaCO_3$-equivalent using the quantity of CO_2 produced on reaction with HCl, the cation exchange capacity (CEC) and exchangeable bases were determined using ammonium acetate (NH_4OAc (1 N, pH 7)) method. The methods used are described in detail in the Soil Survey Laboratory Methods Manual [19].

Free iron oxides (Fe_2O_{3d}) were extracted using the sodium dithionite-citrate (DCB) method [20], amorphous iron oxides (Fe_2O_{3o}) were extracted using ammonium oxalate method [21] and organically bound iron oxides (Fe_2O_{3p}) were extracted using the sodium pyrophosphate method [22]. Total iron was extracted by aqua regia digestion method [23]. The amount of crystalline iron oxides was estimated from the total iron minus amorphous iron oxides minus organically bound iron oxides. All iron concentrations were determined by atomic absorption spectrophotometry, using a Varian SpectrAA 300.

In order to estimate the soil salinity level, in both soil samples, the electrical conductivity of saturation paste extract (EC_e) was determined [24].

2.3. Mineralogical Analysis

Mineralogical analysis was carried out by using an X-Ray Diffractometer of Siemens D 5005 type, with copper tube at 40 KV and 40 mA and graphite monochromatographe, at the Laboratory of Economic Geology and Geochemistry, Department of Geology and Geoenvironment, NKUA. The evaluation of the powder X-Ray Diffraction (XRD) pattern was performed by using Bruker EVA 10.0 program of the DIFFRAC^{plus} software package.

2.4. Measurement of Volumetric Water Content and Apparent Dielectric Permittivity of Soils

In all the experiments conducted, the actual soil water content, θ_m, ranged from oven dry to saturation in equal water content steps, $\Delta\theta = 0.05$ m^3m^{-3}. To this end, air-dried soil samples were thoroughly mixed with different predetermined amounts of fresh water (EC = 0.28 dSm^{-1}) taking care to get homogeneous water content. The obtained soil samples of various water content levels were packed in PVC columns with 10 cm height and 7 cm diameter in small portions and pressed with a 0.15 Kg rubber hammer to achieve homogeneous density distribution in the soil sample. The θ_m and dry bulk density of the soil samples were determined once again at the end of the experiment by weighing and oven drying. In addition, measurements were taken in the oven-dried soil samples ($\theta_m = 0\ m^3m^{-3}$). Using the above methodology, a sufficient number of ε_a, θ_m, and θ readings for the two soils examined were provided. Measurements were obtained by fully inserting vertically the WET and ML2 sensors into the soil samples.

The soil specific calibration of the sensors was performed using the following linear equation between the square root of ε_a and θ_m:

$$\theta_m = a\varepsilon_a^{0.5} - b, \tag{7}$$

where the parameters a and b depend on the soil type. This equation has been widely applied for the calibration of TDR and other dielectric sensors [4,5,7,25–28].

Soil specific values for a and b parameters can be easily determined using two independent pairs of θ_m and ε_a values (two-point calibration, CAL). For simplicity reasons and in order to cover the entire range of possible θ_m values of the studied soil samples, the first ε_a measurement is taken at an oven-dried soil sample (i.e., $\theta_m = 0\ m^3m^{-3}$) and the second ε_a measurement is taken at a saturated soil sample (i.e., $\theta_m = \theta_s$) [28,29]. Optimal values for a and b can be also determined by linear regression between θ_m and $\varepsilon_a^{0.5}$ using all the measured θ_m values (multi-point calibration, CALALL). In this study, soil specific calibration relationships for each soil studied were produced using both the above described methods.

Additionally, corresponding data of another sandy loam soil, referred in the study of Kargas et al. [6], using the same sensors are compared with the recent data received by the soils used in this study. More details about the properties of the soil and the experimental procedure can be found in Kargas et al. [6].

Additionally, a set of experiments where silica sand with bulk density of $1.68\ gcm^{-3}$ was thoroughly mixed with appropriate amounts of soil 1 was carried out. So, sand-soil mixtures with five rates of soil 1 (100% sand, 75% sand and 25% soil 1, 50% sand and 50% soil 1, 25% sand and 75% soil 1, and 100% soil 1 v/v) were made. In each ratio of sand-soil mixture, the relationship between θ_m and ε_a for each sensor, according to the methodology described at the beginning of this subsection, was determined. With these experiments, it is possible to better assess the effect of iron content on ε_a in the case of the sensors tested.

2.5. Performance Evaluation Criteria

To evaluate the calibration equation the Root Mean Square Error (RMSE) performance evaluation criterion was used.

$$RMSE = \sqrt{\frac{\sum\limits_{i=1}^{n}\left(Pred_i - Obs_i\right)^2}{n}}, \tag{8}$$

where $Pred_i$ is the i th predicted value of the modeled parameter, Obs_i is the corresponding observed value, and n is the total number of different observed – predicted values pairs. RMSE is expressed in the same units as the estimated parameter and values close to 0 indicate better performance.

Additionally, the coefficient of determination R^2 was used. R^2 ranges between 0 and 1, where values close to 0 indicate no correlation and values close to 1 indicate a strong correlation.

3. Results and Discussion

3.1. Physical and Chemical Soil Properties

The results of some physical and chemical properties for the two soils used are presented in Table 1.

Table 1. Texture analysis, organic matter (%), content of CaCO$_3$ (%) and cation exchange capacity (CEC) for the two soils used.

Soil Sample	Sand (%)	Silt (%)	Clay (%)	Texture	CaCO$_3$ (%)	CEC (cmol$_+$ Kg^{-1})	Organic Matter (%)
soil 1	62.0	16.0	22.0	Sandy loam	1.47	15.20	2.17
soil 2	66.0	24.0	10.0	Sandy loam	2.1	12.3	0.43

Additionally, the EC$_e$ values were determined 0.565 dSm^{-1} and 0.425 dSm^{-1} for the soil 1 and soil 2, respectively. These values indicated that the soil salinity level is very low and their effect on ε_a can be considered too low.

The content of free and amorphous or non-crystalline iron oxides as well as the organic matter-bound iron oxides and total iron oxides in two soil samples are presented in Table 2.

Table 2. Content of different forms of iron oxides in two sandy loam soils used.

Soil Sample	Content of Different Forms of Iron Oxides (%)			
	Free iron Oxides (Fe$_2$O$_{3d}$)	Amorphous Iron Oxides (Fe$_2$O$_{3o}$)	Organic Matter-Bound Iron Oxides (Fe$_2$O$_{3p}$)	Total Iron
Soil 1	5.97	6.73	0.44	30.12
Soil 2	9.72	1.33	0.024	33.07

From the data showed in Table 2, it appears that both soils have high content of total iron.

In Figure 1 the XRD diagram of soil 1 is presented. The results of the XRD analysis revealed that calcite is the predominant mineral phase and quartz a major to medium constituent. Hematite and goethite occur as medium to minor mineral phases and their co-existence reveals secondary processes (e.g., weathering). Dolomite occurs as a medium to minor crystallic phase (probably attributed to the geological background formations of the area, such as Triassic dolomites and dolomitic limestones). Moreover, illite, chlorite (clinochlore, an iron-rich clay mineral), talc and smectites (montmorillonite and nontronite, an iron-rich smectite) were determined as minor to trace minerals.

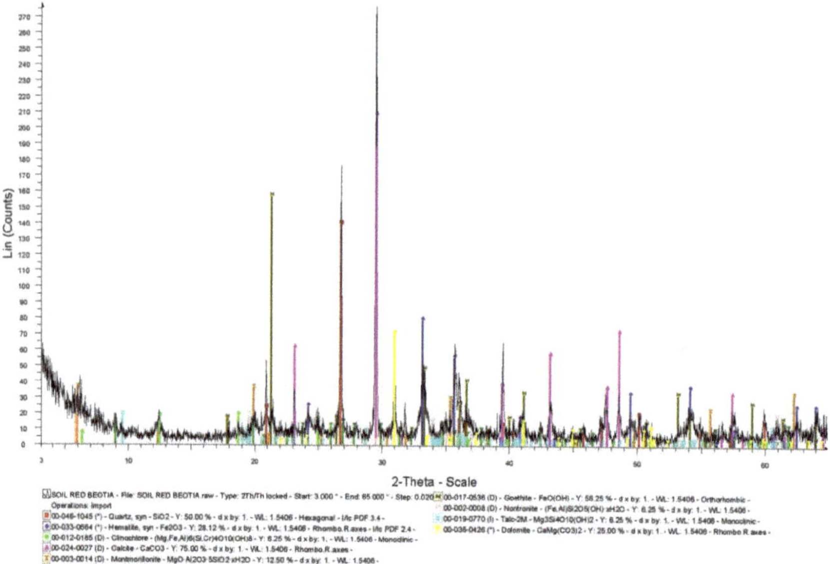

Figure 1. XRD diagram of soil 1.

3.2. Relationship between Actual Soil Water Content and Apparent Dielectric Permittivity

In Figures 2 and 3, the relationships θ_m-ε_a, where ε_a was measured by the WET and ML2 sensors and was also calculated from the Topp equation (Equation (3)) for soils 1 and 2, are presented. The widely accepted Topp equation was used as a reference equation since is independent on soil type, soil temperature and soluble salt content.

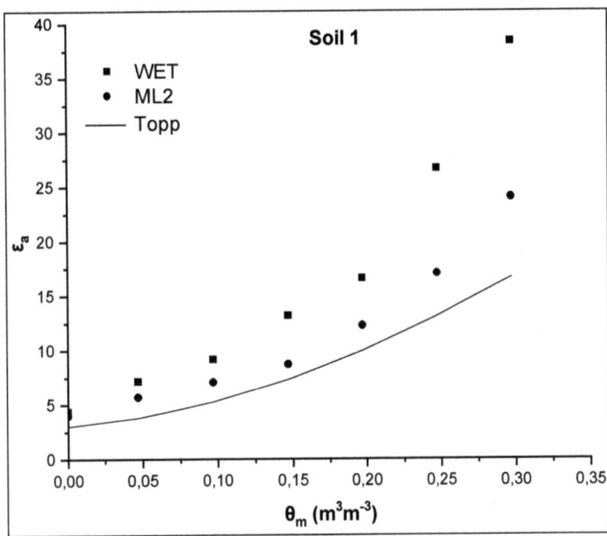

Figure 2. Comparative presentation of relationships θ_m-ε_a obtained by the WET sensor, the ML2 sensor and the TOPP equation for soil 1.

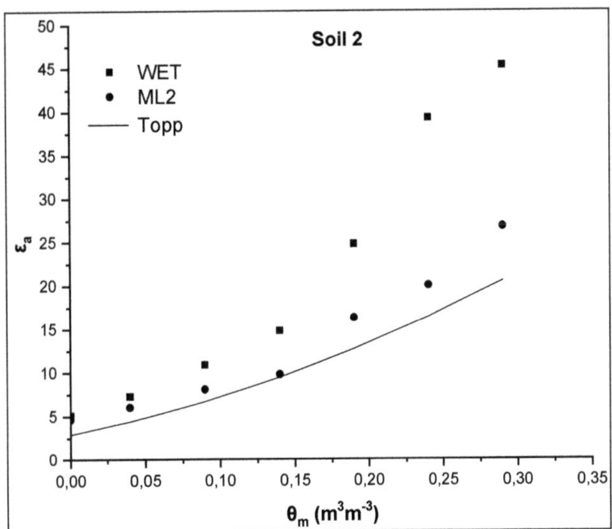

Figure 3. Comparative presentation of relationships θ_m-ε_a obtained by the WET sensor, the ML2 sensor and the TOPP equation for soil 2.

As shown in Figure 2, ε_a values measured by the WET sensor for each θ_m are much higher than those predicted by the Topp equation for the same θ_m value. More specific, for the maximum value of actual soil water content $\theta_m = 0.297$ m^{3}m^{-3}, ε_a values obtained by the WET sensor and the Topp equation are 38.4 and 16.7, respectively. Taking into account the ε_a value 38.4, the maximum θ value obtained by the WET sensor, using the manufacturer calibration (Equation (6)), is 0.435 m^{3}m^{-3}. Also, analogous phenomena, but to a less extent, are observed for the ML2 sensor (Figure 2). The ML2 sensor, which operates at higher frequency, measures lower values of ε_a as anticipated for the same θ_m [6]. For maximum θ_m, the ε_a value obtained by the ML2 sensor is 24, i.e. 14.4 units less than that obtained by the WET sensor, but 7 units greater than that from the Topp equation. Additionally, when the manufacturer calibration is applied using Equation (6) with $\alpha_0 = 1.8$ and $\alpha_1 = 10.1$ for the WET sensor and $\alpha_0 = 1.6$ and $\alpha_1 = 8.4$ for the ML2 sensor, an overestimation of soil water content for $\theta_m > 0.2$ m^{3}m^{-3} and $\theta_m \geq 0$ m^{3}m^{-3} for the ML2 and the WET sensor, respectively, is observed.

In correspondence with soil 1, similar results for the soil 2 are presented in Figure 3. Specifically, ε_a values measured by the WET and the ML2 sensor are 45.3 and 26.8, respectively, for the maximum value of actual soil water content $\theta_m = 0.29$ m^{3}m^{-3}, whereas the ε_a value calculated by the Topp equation is 16.1. Compared to soil 1, it appears that the difference between the ε_a values measured by the two sensors increases and their difference from the predicted ε_a value from the Topp equation increases too as soil water content increases.

The values of ε_a for $\theta_m = 0$ m^{3}m^{-3} are, then, examined. In case of soil 1, ε_a values are 4.4 and 4.02 for WET and ML2, respectively, for $\theta_m = 0$ m^{3}m^{-3}, whereas the corresponding values for soil 2 are 5.1, and 4.65 for WET and ML2, respectively. These values are much higher than those calculated, on the basis of the manufacturer calibration, using Equation (6) for $\theta_m = 0$ m^{3}m^{-3}. From Equation (6) assuming that $\theta_m = 0$ m^{3}m^{-3} then the soil apparent dielectric permittivity is calculated as $\varepsilon_a = \alpha_0{}^2 = 1.8^2 = 3.24$ for the WET sensor and 2.56 for the ML2 sensor. The corresponding value from the Topp equation (Equation (3)) is 3.03.

The high ε_a values measured by the two sensors in the two soils compared to those predicted by the Topp equation may be attributed to the high content of iron and especially to the content of hematite and goethite. Considering the semi-quantitative XRD analysis (Figure 1) and the analysis of different iron forms in soil 1 (Table 2), we can determine that the hematite and goethite content in crystalline phases is approximately 12% and 8%, respectively. Consequently, the sum of the crystalline phases of hematite and goethite is approximately 20%. In particular, the total iron oxides that participate in crystalline phases are: Total iron minus amorphous iron oxides minus organic matter-bound iron oxides (30.12-6.73-0.44 = 22.95%). The remaining percentage of ~3% (2.95%), may be attributed to the contribution of the iron-rich clay minerals (clinochlore and nontronite) as shown by XRD analysis.

Robinson et al. [15] reported that the effect of hematite and goethite on the estimation of ε_a from TDR measurements is clearly less than that of magnetite. However, in the case of capacitance probe (100 MHz) with a rods length of 5 cm and a hematite content of 15%, the effect is considerable (ε_a value rising from 27 in the saturated sand to 44) and even greater in the case of a goethite content of 15% (ε_a value rising from 27 in the saturated sand to 135). In soils 1 and 2 (Figures 2 and 3) which contain hematite and goethite, the behavior of the WET sensor, that operates at 20 MHz and has a 6.8 cm rods length, is similar.

Thus, in the case of low-frequency FDR sensors (i.e. WET), the effect of high soil iron content and particular high content of hematite and goethite may be significant on the soil apparent dielectric permittivity measurement. However, from the present work, we cannot distinguish the particular contribution of hematite and goethite to the end result. Additionally, it could be noted the possible contribution of the clay minerals, such as smectite and chlorite, as it is shown from XRD analysis, on the high ε_a values. This issue needs further investigation to accurately assess the contribution of clay minerals.

Furthermore, the measurements of the soil bulk electrical conductivity (σ_b) using the WET sensor show that the high values of ε_a cannot be attributed to the salinity level of the soil samples since the maximum values of σ_b are 0.46 and 0.5 dSm^{-1} for soil 1 and soil 2, respectively. These low σ_b values in combination with the low EC$_e$ values (0.565 dSm^{-1} and 0.425 dSm^{-1} for the soil 1 and soil 2, respectively) showed that the effect of soil salinity may be considered negligible. It is mentioned that in this range of σ_b, none of the sensors was sensitive as the above σ_b values are much lower than the threshold 2 dSm^{-1} [6].

In order to enhance the results, the ε_a values measured by the two sensors and those ones of the sandy loam soil studied by Kargas et al. [6] with completely different mineralogical composition (most common constituent was illite with chlorite much less prominent), low soil salinity (with EC$_e$ = 0.655 dSm^{-1}) and insignificant iron content (0.9%) were compared. In this case, the difference between the two sensors is approximately 4 dielectric units, at the maximum θ_m, and the difference between the WET sensor and the Topp equation is 5 dielectric units, while the difference between the ML2 sensor and the Topp equation is negligible (Figure 4). The values of ε_a, at the maximum θ_m, are 23 and 19 for the WET and the ML2, respectively, i.e., much lower than the values measured in Soil 1 and Soil 2, especially for the WET sensor which operates at low frequencies.

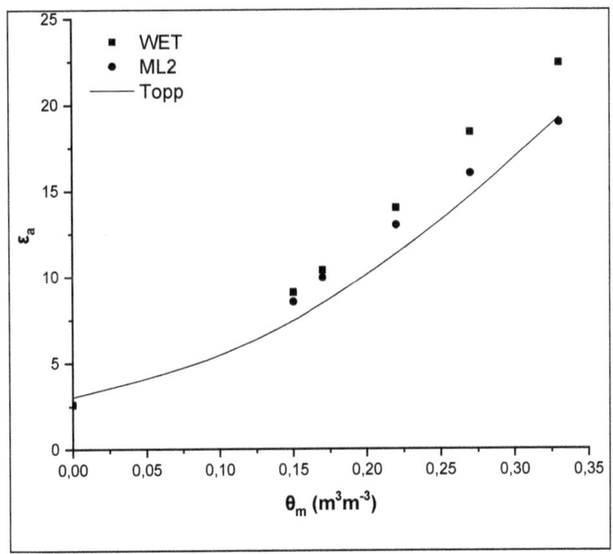

Figure 4. Comparative presentation of the relationships θ_m-ε_a obtained by the WET sensor, the ML2 sensor and the Topp equation for a sandy loam soil (SL) studied by Kargas et al. (2014) [6] with low iron content.

From these results, it is confirmed that, due to the iron content difference, it is observed a great difference in ε_a between the sandy loam soil studied by Kargas et al. [6] and the two sandy loam soils examined in the present study. The high θ values obtained, in both soils studied, by the WET and ML2 sensors taking into consideration the manufacturer calibration, highlight that a specific calibration of the sensors is required for these soils. Results, also, show the influence of the soil mineralogical composition. Overall, it can be said that the soil texture alone is not an effective way to categorize soils in relation to sensors calibration, as in the case of the manufacturer calibration of WET and ML2 sensors.

3.3. Soil Specific Calibration

During the specific calibration process, the linearity of θ_m-$\varepsilon_a^{0.5}$ relationship was investigated. In Figure 5, a comparative presentation between the θ_m-$\varepsilon_a^{0.5}$ relationships obtained by the WET and ML2 sensors for soil 1 is presented, together with the respective manufacturer calibration, while in Table 3, the coefficients of these linear relationships (CALALL) are given.

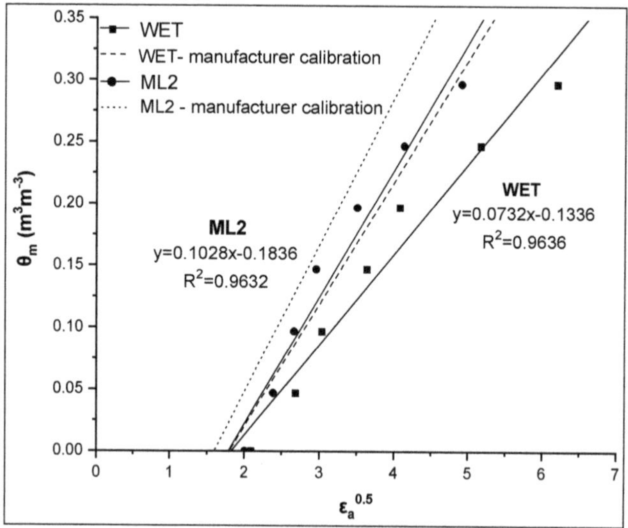

Figure 5. Comparative presentation of the relationships θ_m-$\varepsilon_a^{0.5}$ obtained by the WET and ML2 sensors for soil 1. The corresponding relationships using the manufacturer calibration are also given for both sensors.

Table 3. Coefficients (slope (a) and intercept (-b)) of the linear relationship θ_m-$\varepsilon_a^{0.5}$ obtained by the WET and ML2 sensors (CALALL) for the two sandy loam soils studied (Soil 1 and Soil 2). Additionally, the coefficients of a Sandy Loam (SL) soil studied by Kargas et al. [6] are presented.

Soil	WET		ML2	
	a	b	a	b
Soil 1	0.073	0.133	0.102	0.183
Soil 2	0.060	0.116	0.093	0.182
SL	0.104	0.173	0.116	0.195

As shown in Figure 5, the relationship θ_m-$\varepsilon_a^{0.5}$ is strongly linear for both sensors used (high value of R^2). However, in the case of the WET sensor, the coefficients of this linear relationship (slope and intercept) are smaller than those obtained by the Topp equation (i.e. slope = 0.115, intercept = −0.176), while in the case of the ML2 sensor, there is a difference mainly in the slope value. From RMSE values, it appears that the estimation of soil water content by the CALALL method is significantly improved using the specific calibration compared to the manufacturer calibration. Specifically, the RMSE values are 0.018 and 0.019 m^3m^{-3} for the WET and the ML2 sensor, respectively, while the corresponding values taking into account the manufacturer calibration were 0.067 and 0.051 m^3m^{-3}, respectively. If we consider the RMSE value of 0.03 m^3m^{-3} as a threshold to achieve a reliable calibration then it seems that the CALALL method is quite effective.

Similar findings for soil 2 are shown in Figure 6 and Table 3 (CALALL). In the case of soil 2, RMSE values by the CALALL method are 0.015 and 0.014 for the WET and the ML2 sensor, respectively. The corresponding values for manufacturer calibration were 0.123 and 0.086 m^3m^{-3}, respectively.

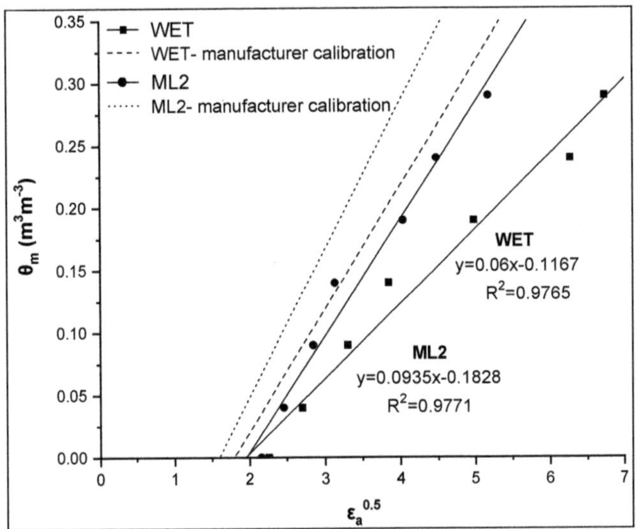

Figure 6. Comparative presentation of the relationships θ_m-$\varepsilon_a^{0.5}$ obtained by the WET and ML2 sensors for soil 2. The corresponding relationships using the manufacturer calibration are also given for both sensors.

Therefore, it is possible to overcome the problem of soil water content overestimation, due to the overestimation of ε_a values for this particular category of soils, using the specific calibration (CALALL) of sensors. Additionally, the two-point calibration method (CAL) appears to be effective, since RMSE values for soil 1 are 0.032 and 0.029 m^3m^{-3} for WET and ML2, respectively, whereas for soil 2 are 0.019 and 0.023 m^3m^{-3}, respectively. Although, these values are higher than those of the CALALL method, they are less than the threshold value 0.03 m^3m^{-3}, with exception of soil 1 values for the WET sensor.

3.4. Sand-Soil Ratio

Figure 7 shows the relationship θ_m-ε_a for different sand-soil 1 ratios (v/v) in soil mixture studied using the WET sensor. The results showed the effect of soil 1 content on the ε_a value even in the case of a small percentage of soil 1 addition in soil mixture, i.e., 25% by volume. More specific, between the pure sand (100% sand) and the soil 1 (100% soil 1) there is a difference of 25 dielectric units at the maximum soil water content. Each 25% increase of the soil 1 in soil mixture leads to an increase of approximately 7 dielectric units, except in the case between the percentages 75% and 100% where the increase is less. Similar results were obtained by the ML2 sensor, where the increase of ε_a values is approximately 4 dielectric units for each percentage increase of the soil 1 in soil mixture by 25% (data not shown).

Therefore, there is a strong positive correlation between the soil 1 content and ε_a. The smallest value is obtained in 100% sand and as the soil 1 percentage increases in the soil mixture the ε_a increases remarkably (Figure 7).

This underscores the need to account for the effect of soil iron content on sensor readings. This can be achieved by developing specific calibration equations as presented in Table 3.

Additionally, the strong linear correlation between ε_a and the soil 1 percentage in the soil mixture studied at various soil water contents and especially at the maximum actual soil water content value, i.e., $\theta_m = 0.297$ m^3m^{-3}, ($R^2 = 0.9945$) as shown in Figure 8 for the WET sensor, is noteworthy. This study's findings clearly show that the addition of soil with high iron content results in higher apparent dielectric permittivity and thus, cause an overestimation of θ.

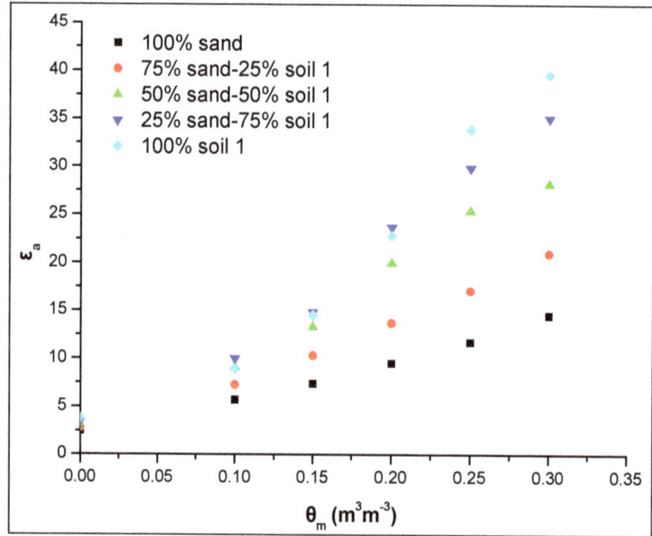

Figure 7. Relationship between the apparent dielectric permittivity ε_a and the actual soil water content θ_m for five different ratios of the sand-soil mixture studied using the WET sensor.

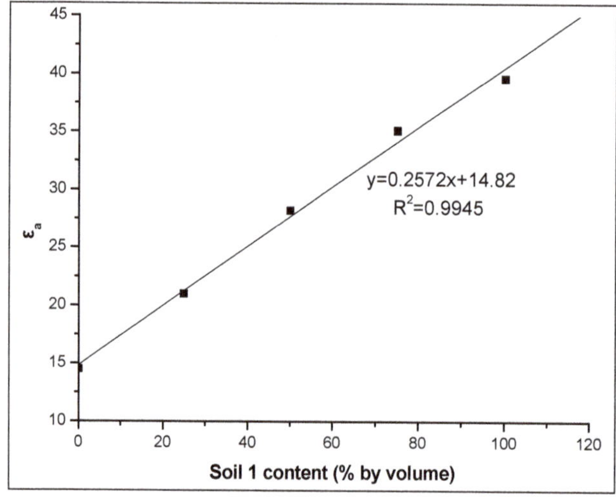

Figure 8. Relationship between the apparent dielectric permittivity ε_a and the percentage of soil 1 in the soil mixture studied at the maximum actual soil water content using the WET sensor.

4. Conclusions

From the experiments conducted by using two dielectric sensors on two sandy loam soils, it is clearly shown that high iron content leads to high apparent dielectric permittivity values. Higher values of ε_a were determined when the measurements were made by the WET sensor, due to the fact that it operates at a lower frequency than the ML2. The high values of ε_a lead to an overestimation of soil water content in the case where the manufacturer calibration of the sensors is used. Soil water content near saturation is overestimated up to 60% using the WET sensor.

The problem of soil water content overestimation can be overcome by specific calibration (especially CALALL) of the sensors for the soils used. Even in these cases of soils, the relation θ_m-$\varepsilon_a^{0.5}$ is linear, and the coefficients of this relationship for each soil can be easily determined. The coefficients for these soils are different from those obtained by the Topp equation and those proposed by the manufacturer for inorganic soils.

Further investigation is needed to distinguish the particular contribution of hematite and goethite to the higher values of the apparent dielectric permittivity.

Overall, our findings confirmed that iron content significantly affects the soil apparent dielectric permittivity and thus further investigation is needed under different soil types and environmental conditions.

Author Contributions: Conceptualization, G.K., P.L., M.A. and N.M.; Formal analysis, G.K., P.L., M.A. and N.M.; Methodology, G.K., P.L., M.A. and N.M.; Writing—review & editing, G.K., P.L., M.A. and N.M. All authors have read and agreed to the published version of the manuscript.

Funding: This research received no external funding.

Conflicts of Interest: The authors declare no conflict of interest.

References

1. Topp, G.C.; Davis, J.L.; Annan, A.P. Electromagnetic determination of soil water content: Measurements in coaxial transmission lines. *Water Resour. Res.* **1980**, *16*, 574–582. [CrossRef]
2. Kargas, G.; Kerkides, P. Water content determination in mineral and organic porous media by ML2 theta probe. *Irrig. Drain.* **2008**, *57*, 435–449. [CrossRef]
3. Topp, G.C.; Reynolds, W. Time domain reflectometry: A seminal technique for measuring mass and energy in soil. *Soil Tillage Res.* **1998**, *47*, 125–132. [CrossRef]
4. Ferre, P.A.; Topp, G.C. Time domain reflectometry. In *Methods of Soil Analysis; Part 4-Physical Methods*; Soil Science Society of America, Inc.: Madison, WI, USA, 2002; pp. 434–446.
5. Kargas, G.; Kerkides, P.; Seyfried, M.; Sgoumbopoulou, A. WET Sensor Performance in Organic and Inorganic Media with Heterogeneous Moisture Distribution. *Soil Sci. Soc. Am. J.* **2011**, *75*, 1244–1252. [CrossRef]
6. Kargas, G.; Kerkides, P.; Seyfried, M. Response of Three Soil Water Sensors to Variable Solution Electrical Conductivity in Different Soils. *Vadose Zone J.* **2014**, *13*. [CrossRef]
7. Kargas, G.; Soulis, K. Performance evaluation of a recently developed soil water content, dielectric permittivity, and bulk electrical conductivity electromagnetic sensor. *Agric. Water Manag.* **2019**, *213*, 568–579. [CrossRef]
8. Hamed, Y.; Samy, G.; Persson, M. Evaluation of the WET sensor compared to time domain reflectometry. *Hydrol. Sci. J.* **2006**, *51*, 671–681. [CrossRef]
9. Robinson, D.A.; Jones, S.B.; Wraith, J.M.; Or, D.; Friedman, S.P. A review of advances in dielectric and electrical conductivity measurement in soils using time domain reflectometry. *Vadose Zone J.* **2003**, *2*, 444–475. [CrossRef]
10. Delta-T Devices. *User Manual for the WET Sensor (Type WET-2)*; Delta-T Device Ltd.: Cambridge, UK, 2007.
11. Delta-T Device. *User Manual for the MML2 Sensor*; Delta-T Device Ltd.: Cambridge, UK, 1999.
12. Inoue, M.; Ahmed, B.O.; Saito, T.; Irshad, M. Comparison of Twelve Dielectric Moisture Probes for Soil Water Measurement under Saline Conditions. *Am. J. Environ. Sci.* **2008**, *4*, 367–372. [CrossRef]
13. Campbell, J.E. Dielectric Properties and Influence of Conductivity in Soils at One to Fifty Megahertz. *Soil Sci. Soc. Am. J.* **1990**, *54*, 332–341. [CrossRef]
14. Roth, C.H.; Malicki, M.A.; Plagge, R. Empirical evaluation of the relationship between soil dielectric constant and volumetric water content as the basis for calibrating soil moisture measurements by TDR. *J. Soil Sci.* **1992**, *43*, 1–13. [CrossRef]
15. Robinson, D.A.; Bell, J.; Batchelor, C. Influence of iron minerals on the determination of soil water content using dielectric techniques. *J. Hydrol.* **1994**, *161*, 169–180. [CrossRef]
16. Van Dam, R.; Schlager, W.; Dekkers, M.J.; Huisman, J.A. Iron oxides as a cause of GPR reflections. *Geophys.* **2002**, *67*, 536–545. [CrossRef]
17. Pettinelli, E.; Vannaroni, G.; Cereti, A.; Pisani, A.R.; Paolucci, F.; Del Vento, D.; Dolfi, D.; Riccioli, S.; Bella, F. Laboratory investigations into the electromagnetic properties of magnetite/silica mixtures as Martian soil simulants. *J. Geophys. Res. Space Phys.* **2005**, *110*. [CrossRef]

18. Kargas, G.; Kerkides, P. Performance of the theta probe ML2 in the presence of nonuniform soil water profiles. *Soil Tillage Res.* **2009**, *103*, 425–432. [CrossRef]

19. Soil Survey Laboratory Methods Manual. *Soil Survey Investigations Report No.42, Version 4.0, November 2004*; United States Department of Agriculture, Natural Resources Conservation Service: Lincoln, NE, USA, 2004.

20. Mehra, O.P. Iron Oxide Removal from Soils and Clays by a Dithionite-Citrate System Buffered with Sodium Bicarbonate. *Clays Clay Miner.* **1958**, *7*, 317–327. [CrossRef]

21. Blakemore, L.C.; Searle, P.L.; Daly, B.K. *Methods for Chemical Analysis of Soils*; N.Z. Soil Bureau, Scientific Report 80; N.Z. Soil Bureau: Lower Hutt, New Zealand, 1987.

22. Bascomb, C.L. Distribution of pyrophosphate-extractable iron and organic carbon in soils of various groups. *J. Soil Sci.* **1968**, *19*, 251–268. [CrossRef]

23. Chen, M.; Ma, L.Q. Comparison of Three Aqua Regia Digestion Methods for Twenty Florida Soils. *Soil Sci. Soc. Am. J.* **2001**, *65*, 491–499. [CrossRef]

24. U.S. Salinity Laboratory Staff. *Diagnosis and Improvement of Saline and Alkali Soils*; U.S. Salinity Laboratory Office: Washington, DC, USA, 1954.

25. Ledieu, J.; De Ridder, P.; De Clerck, P.; Dautrebande, S. A method of measuring soil moisture by time-domain reflectometry. *J. Hydrol.* **1986**, *88*, 319–328. [CrossRef]

26. White, I.; Knight, J.; Zegelin, S.; Topp, G. Comments on 'Considerations on the use of time-domain reflectometry (TDR) for measuring soil water content' by W.R. Whalley. *Eur. J. Soil Sci.* **1994**, *45*, 503–508. [CrossRef]

27. Spaans, E.J.A.; Baker, J.M. The Soil Freezing Characteristic: Its Measurement and Similarity to the Soil Moisture Characteristic. *Soil Sci. Soc. Am. J.* **1996**, *60*, 13–19. [CrossRef]

28. Kargas, G.; Soulis, K. Performance Analysis and Calibration of a New Low-Cost Capacitance Soil Moisture Sensor. *J. Irrig. Drain. Eng.* **2012**, *138*, 632–641. [CrossRef]

29. Seyfried, M.S.; Murdock, M.D. Measurement of soil water content with a 50 MHz soil dielectric sensor. *Soil Sci. Soc. Amer. J.* **2004**, *68*, 394–403. [CrossRef]

© 2020 by the authors. Licensee MDPI, Basel, Switzerland. This article is an open access article distributed under the terms and conditions of the Creative Commons Attribution (CC BY) license (http://creativecommons.org/licenses/by/4.0/).

Article

Optimization of Spring Wheat Irrigation Schedule in Shallow Groundwater Area of Jiefangzha Region in Hetao Irrigation District

Zhigong Peng [1,2], Baozhong Zhang [1,2,*], Jiabing Cai [1,2], Zheng Wei [1,2,*], He Chen [1,2] and Yu Liu [1,2]

[1] State Key Laboratory of Simulation and Regulation of Water Cycle in River Basin, China Institute of Water Resources and Hydropower Research, Beijing 100038, China; pengzhg@iwhr.com (Z.P.); caijb@iwhr.com (J.C.); chenhe@iwhr.com (H.C.); liuyu@iwhr.com (Y.L.)

[2] National Center of Efficient Irrigation Engineering and Technology Research-Beijing, Beijing 100048, China

* Correspondence: zhangbaozhong333@163.com (B.Z.); weizheng@iwhr.com (Z.W.); Tel.: +86-10-6878-6510 (B.Z.); +86-10-6878-5226 (Z.W.)

Received: 28 September 2019; Accepted: 10 December 2019; Published: 13 December 2019

Abstract: Due to the large spatial variation of groundwater depth, it is very difficult to determine suitable irrigation schedules for crops in shallow groundwater area. A zoning optimization method of irrigation schedule is proposed here, which can solve the problem of the connection between suitable irrigation schedules and different groundwater depths in shallow groundwater areas. The main results include: (1) Taking the annual mean groundwater depth 2.5 m as the dividing line, the shallow groundwater areas were categorized into two irrigation schedule zones. (2) On the principle of maximizing the yield, the optimized irrigation schedule for spring wheat in each zone was obtained. When the groundwater depth was greater than 2.5 m, two rounds of irrigation were chosen at the tillering–shooting stage and the shooting–heading stage with the irrigation quota at 300 mm. When the groundwater depth was less than 2.5 m, two rounds of irrigation were chosen at the tillering–shooting stage, and one round at the shooting–heading stage, with the irrigation quota at 240 mm. The main water-saving effect of the optimized irrigation schedule is that the yield, the soil water use rate, and the water use productivity increased, while the irrigation amount and the ineffective seepage decreased.

Keywords: crop model; water consumption; yield; water production function; irrigation schedule optimization

---

1. Introduction

In a shallow groundwater area, the groundwater is supplied to the aeration zone through capillary rise becoming soil water available to the crops. The interaction between soil water and groundwater varies due to the depth of groundwater [1–4]. For the sake of greater water economy, crop yield, and seeking the greatest advantage from the regulating effect of groundwater in soil water, scholars are particularly interested in the impact of different groundwater depths on crop growth. Kong et al. [5] studied the effect of different groundwater depths on crop growth using a lysimeter, finding that a depth of 1.5–2.5 m was conducive to crop growth, and when this depth was more than 2.0 m, the existent irrigation schedule was unable to meet the normal growth of crops. Kruse et al. [6] pointed out that in the areas with shallow groundwater depth, the groundwater recharge affected the water and the biological and chemical processes of the soil–plant–atmosphere continuum, and if no irrigation was provided, the optimal groundwater depth for winter wheat was about 1.5 m. Wang et al. [7] studied the effect of different groundwater depths on crop growth, showing that different groundwater depths led to differences in crop root distribution, which in turn affected the crops' water-yield response mechanism. Zhang [8] using a lysimeter, studied the drought crops' groundwater utilization, suggesting that in the suitable groundwater depth, the groundwater used by drought crops accounted for 50% to 70% of the evapotranspiration. Yang et al. [9] using the HYDRUS software, simulated the influence of different groundwater depths on the irrigation quota of mulched, drip-irrigated cotton, finding that the drip irrigation quota was 330 mm, 450 mm, or 550 mm with the groundwater depth respectively at 1.5 m, 2.0 m, and 3.0 m. Zhuang et al. [10] studied the recharge effect of different groundwater depths on the cotton root layer, pointing out that when the groundwater depth did not exceed 2 m, the cotton irrigation schedule should be developed with the consideration of the groundwater recharge. Wang et al. [11] studied the spring wheat recharge modes under different groundwater depths, showing that the recharge was the largest at the groundwater depth of 1.0 m, and there was basically no replenishment with the groundwater depth at 3.0 m or greater. Liu et al. [12] technically supported by a lysimeter with controlled groundwater depth, determined the deficit irrigation schedule for crops under different groundwater depths. Relevant researchers pointed out that groundwater action was particularly critical in the analysis of the soil–crop–atmosphere system water balance in an arid oasis [13–18]. The deficit irrigation schedule in shallow groundwater areas could improve groundwater utilization but limited the influence on yield [19–21]. Karimov et al. [22] pointed out that a shallower groundwater depth promoted phreatic evaporation.

In summary, the proportion of phreatic evaporation varies notably with groundwater depth. To be rational, an irrigation schedule should fully consider the groundwater recharge under different groundwater depths. Previous studies on the effect of different groundwater depths on crop growth were mainly based on the controlled groundwater table by a lysimeter, with the groundwater table remaining unchanged during the whole crop growth period, which is not in line with the actual situation, because there is significantly daily variation in the groundwater table throughout the crop growth period. Therefore, the studies based on controlled groundwater table can hardly represent the actual change of groundwater depth throughout the crop growth period. The studies on the effect of different groundwater depths on crop growth, irrigation amount, and irrigation schedule optimization are mostly based on experiment stations; however, in shallow groundwater depth areas the groundwater table varies greatly from place to place. Therefore, the problem of how to apply the experimental results to a large expanse of areas in urgent need of a solution.

Hetao Irrigation District is the largest gravity irrigation district by water diverted from the Yellow River. According to the overall water allocation plan of the Yellow River watershed, the quota of water diversion to Hetao Irrigation District has decreased from 5.18×10^9 m^3 to 4.00×10^9 m^3. This ever-decreasing diversion will gravely affect the grain production in the irrigation district, making the conflict between supply and demand even more serious [23,24]. After the implementation of water-saving projects in the Hetao Irrigation District, the amount of water diversion for agricultural purposes has been cut notably. The result is that the groundwater table has been falling year on year [25]. Li et al. [26] pointed out that the spatial variation of groundwater depth was great in the Jiefangzha Region, and in the well irrigation area, the groundwater table was of a funnel shape with a groundwater depth more than 2.5 m, and in some localities, the groundwater table exceeded 4.5 m. It can be seen that with dwindling water diversion from the Yellow River and the growing well irrigation area, the spatial difference in groundwater depth is increasing.

The findings of previous studies on the crop irrigation schedule in Hetao Irrigation District were based on groundwater table at experiment stations in specific years. The results from experiment stations can hardly reflect the great difference in groundwater depth throughout the irrigation district, and so the application of related findings to a larger area has great limitations. Spring wheat is one of the main grain crops in Hetao Irrigation District, and wheat production plays an important role in grain production in this district. As spring wheat in Hetao Irrigation District grows in the dry season, irrigation is the key to its high yield. In Hetao Irrigation District, the net irrigation quota of spring wheat has been cut to about 300 mm. In shallow groundwater depth areas, it is difficult to maximize the use of the soil water in the soil and thus the water use efficiency is low. In areas with greater groundwater depth, the groundwater recharge is reduced, resulting in the water deficit during certain growth stages.

Compared with the field experiments, studying crop water consumption characteristics based on models has benefits such as the freedom from geographical restrictions, time and financial efficiency, and additional system observables. In addition to the above, it is also possible to remove some interference factors, thus helpful to expose some behaviors among variables. Therefore, technically built on a verified crop growth simulation model, this study investigates the water-yield response mechanism of spring wheat for different groundwater depths, and constructs a spring wheat water production function for each zone. From the above information, an optimization method of zoning irrigation schedule is developed, which solves the problem of groundwater spatial variability in shallow groundwater areas. It is hoped that this study may provide some useful reference for the optimization of irrigation schedules in shallow groundwater areas.

2. Materials and Methods

2.1. Study Area

The experiment was carried out from March 2015 to July 2016 in the Jiefangzha Region of Hetao Irrigation District, Inner Mongolia. The Jiefangzha Region is at N 40°32′–N 41°11′, E 106°51′–107°23′, and its elevation varies between 1030–1046 m. Most of the irrigation area is located within the jurisdiction of Hangjinhouqi of Inner Mongolia Autonomous Region. The Jiefangzha Region, with a controlled area of 21.57×10^4 hm^2 and an irrigated area of 14.21×10^4 hm^2, is the second largest in Hetao Irrigation District. This irrigation area has a comparatively flat terrain, and the overall terrain, high in the southwest and low in the northeast, has an average slope of about 0.02%. The Jiefangzha Region is featured by the arid or semiarid climate. The average annual precipitation and evaporation from a free water surface are 140 mm and 2096 mm respectively, and the annual average temperature is 9 °C. The average annual sunshine hours are 3181 h, the frost-free period is 130–150 d, and the annual average groundwater depth is 1.86 m. According to the American soil classification system, the soil of this irrigation area is dominated by silt loam. Table 1 summarizes the soil's physical properties in the study area.

Table 1. Soil physical properties in the study area.

Depth (cm)	Dry Bulk Density (g/cm^3)	Saturated Moisture Content (m^3/m^3)	Field Capacity (m^3/m^3)	Wilting Point (m^3/m^3)
0–10	1.45	0.46	0.36	0.09
10–20	1.40	0.47	0.38	0.09
20–40	1.34	0.49	0.41	0.09
40–60	1.38	0.48	0.41	0.08
60–100	1.34	0.50	0.42	0.09

2.2. Design of the Experiment

The spring wheat variety tested was Yongliang No. 4. The spring wheat in 2015 was sowed and harvested on 19 March and 19 July, respectively, with the precipitation during the growth period being 61 mm. The spring wheat in 2016 was sowed and harvested on 14 March and 18 July, respectively, with the precipitation during the growth period being 55 mm. According to the water distribution of the irrigation region in previous years, a total of four rounds of irrigation are made throughout the spring wheat growth period. However, at the time of the fourth irrigation, the spring wheat had already been at the ripening stage, and therefore this irrigation contributed little to wheat yield. For this reason, local farmers rarely make the fourth irrigation. To improve the water productivity of spring wheat, the experiment included the first three rounds of irrigation only.

Five treatments were provided, as shown in Table 2, with three replications. Each experiment plot area was 20 m^2. In order to preclude lateral permeability between the plots, each plot was fringed with a 1 m-wide protection row. According to the soil moisture of each experiment treatment at the time of water distribution, the irrigation quota was estimated such that the irrigation upper limit should not exceed the field capacity. Each experiment plot area was irrigated by pumping from the canal. The irrigation volume of the experiment plots was measured by water meters. The field management practice, such as sowing, fertilizing, and farming, for each experiment plot was the same as that of the local farmers.

Table 2. Experiment treatments.

Treatment	Tillering–Shooting	Shooting–Heading	Heading–Filling	Irrigation Quota (mm)	
				2015	2016
T1				-	-
T2		√		100	100
T3	√	√		160	160
T4	√		√	160	135
T5	√	√	√	260	235

Note: "√" means irrigation at this growing stage.

2.3. Data Observation

The relevant meteorological data include solar radiation, wind speed, temperature, atmospheric humidity, and rainfall, all taken from the Hangjinhouqi National Meteorological Station, close to the study area about 1 km. The Penman–Monteith formula, recommended by FAO, was utilized to estimate the reference crop evapotranspiration (ET$_0$) based on the longitude, latitude, and altitude of the weather station [27]. From the Shahaoqu Experimental Station in this irrigation area, the study area groundwater table data of 57 observation wells from 1990–2016 were collected. A groundwater table distribution map was generated using the inverse distance weighting interpolation. At the same time, the groundwater table monitoring wells were also installed in the experimental site, which were read once every 2 or 3 days during the study period. The soil moisture content was determined by the oven drying method. Samples were taken from each plot at an interval of 5 days, and extra measurements were taken before and after rainfall and irrigation. Sampling depths were at 0~20 cm, 20~40 cm, 40~60 cm, 60~80 cm, and 80~100 cm. Upon the harvest, the yield of spring wheat was evaluated. For this purpose, a representative 1 m^2 quadrat was chosen from each experiment plot to determine the grain yield after natural air drying. The temperature, precipitation, reference crop evapotranspiration, and groundwater table change in the experiment plots throughout the experiment period are as shown in Figure 1. The interannual variation of the groundwater table in this irrigation area is shown in Figure 2.

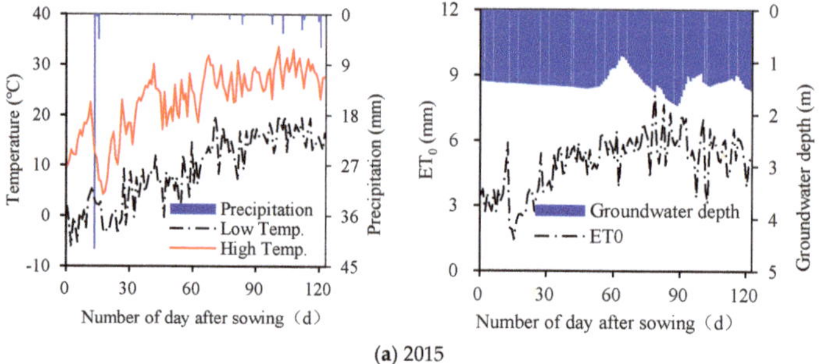

(a) 2015

Figure 1. *Cont.*

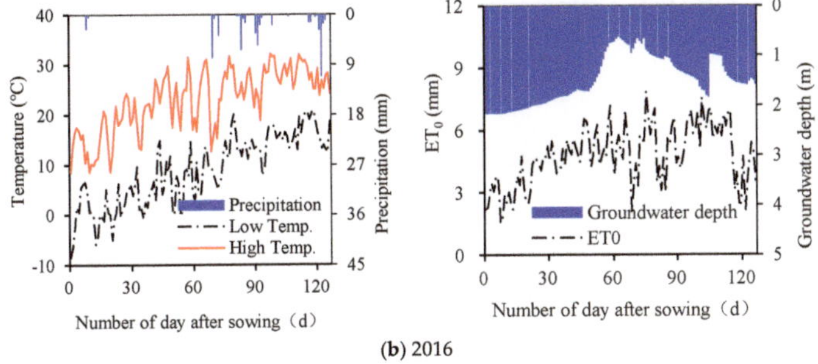

(b) 2016

Figure 1. Meteorological data and groundwater table of the study area during the growth period of spring wheat.

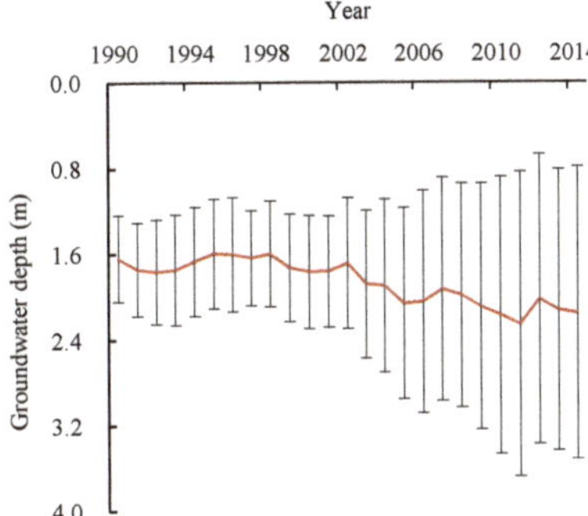

Figure 2. Interannual variation of groundwater table in Jiefangzha Region.

2.4. The Aquacrop Model

The evapotranspiration is divided into two parts by the model [28–31]: Evaporation and transpiration. In order to separate the evaporation, the transpiration was estimated based on the variation of the crop canopy ground cover instead of leaf area index in the whole growth period. Crop yield is calculated based on the biomass on the ground and the harvest index. Based on the difference in the influence mechanism of environment on biomass and on harvest index, the effects of environmental stresses on biomass and harvest index were distinguished. By limiting canopy stretching, accelerating canopy senescence, controlling stomatal closure, and regulating harvest index after the start of reproductive growth, the soil water stresses on crop growth were further refined. From this basis, the crop yields under different irrigation schedules were simulated. The input data of the crop's water-yield response mechanism simulation included crop species, meteorology, soil, groundwater, and irrigation schedule, field management, and initial conditions.

2.5. Model Verification

The input database for crop model consists of crop growth data, meteorological data, soil properties, irrigation schedules, and field management data. For the study area, the soil properties and field management data have remained unchanged during the 2-year experiment. The measured data such as crop growth data, meteorological data and irrigation schedules and so on from the 2015 spring wheat were used to calibrate the model, and those from the 2016 spring wheat were used to verify the model. Soil moisture and yield were used to verify the model parameters. The major parameters of the Aquacrop model for simulating the growth of spring wheat in the Hetao Irrigation District are as shown in Table 3.

Table 3. Some parameters of spring wheat for the crop growth simulation model.

Parameter	Default	Calibrated
Cutoff temperature (°C)	26	26
Crop coefficient	1.10	1.05
Upper and lower thresholds of soil water depletion coefficient	0.20~0.65	0.15~0.35
Shape factor for water stress coefficient for canopy expansion	5.0	5.0
Upper stomatal control limit coefficient of soil stress	0.65	0.35
Soil water depletion fraction for stomatal control-upper threshold	2.5	2.5
Canopy growth coefficient	0.04901	0.07600
Canopy decline coefficient	0.07179	0.18506
Maximum canopy cover in fraction soil cover	0.96	0.98
Minimum effective rooting depth (m)	0.30	0.30
Maximum effective rooting depth (m)	1.50	0.90
Normalized water productivity (g/m^2)	15.0	19.7
Harvest index (%)	48	48
Number of plants per hectare	4,500,000	6,500,000

In the verification process, the degree of agreement between the simulated and the observed value was evaluated by root mean square error (RMSE), mean absolute error (MAE), mean relative error (MBE), and the Nash efficiency coefficient (EF). RMSE and MAE is used to test the unbiasedness of the model, resulting in that the lower their values, the less biased the model, and thus the more accurate the simulation. The EF is a kind of relative error index, also a dimensionless model evaluation index. When taking a value close to 1, the model was believed to have high credibility. A value close to zero suggests that, though the simulation result is generally credible, the simulation process involves larger errors. When the MBE is greater than 0, the simulation result is believed to be on the greater side; otherwise, on the smaller side. The model evaluation indices are determined by [32–34]:

$$RMSE = \sqrt{\frac{1}{n}\sum_{i=1}^{n}(M_i - Q_i)^2} \tag{1}$$

$$MAE = \frac{1}{n}\sum_{i=1}^{n}|M_i - O_i| \tag{2}$$

$$MBE = \frac{1}{n}\sum_{i=1}^{n}(M_i - Q_i) \tag{3}$$

$$EF = 1.0 - \frac{\sum_{i=1}^{n}(O_i - M_i)^2}{\sum_{i=1}^{n}\left(O_i - \overline{O}\right)^2} \tag{4}$$

where, O_i, M_i, and $\overline{O}$ stand for the measured value, simulated value, measured mean value; n is the times of measurement

2.6. Scenarios

2.6.1. Determination of the Typical Year

The precipitation data in the study area from 1961 to 2014 were analyzed, finding that the average annual precipitation during the spring wheat growth period was 61 mm; the year closest to the typical annual precipitation was 2013, with a precipitation of 58 mm.

2.6.2. Determination of Groundwater Depth

In light of the gentle terrain in the Jiefangzha Region, the year-by-year groundwater depth data of the 57 monitoring wells from 1990 to 2015 were interpolated using the inverse distance weighting method, which indicated the mean annual groundwater depth in this area was 1.6–2.3 m and the depth exceeded 2.2 m, in 2010, 2011, and 2014.

According to the phreatic evaporation data of the Shahaoqu Experimental Station, once the groundwater depth exceeded 2.5 m, the phreatic evaporation was significantly reduced. Groundwater depth was closely related to grain yield, for which the shallower the depth, the more serious the soil salinization was and the lower the grain yield was [35]. Still, relevant studies showed that when groundwater depth exceeded 2.5 m, the ecological environment in an arid irrigation district might be adversely affected [36].

Without compromising the ecological safety, and for the sake of preventing soil salinization and minimizing water diversion from the Yellow River, the average annual groundwater depth was taken to be 2.5 m for the future scenario. The interannual spatial variation and the intraanual difference of groundwater table were based on the mean value of 2010, 2011, and 2014. For the future scenario, the spatial distribution of groundwater depth and the zoning of the irrigation schedule are shown in Figure 3. With a groundwater depth of 2.5 m as the divide, the area was divided into zones with significant influence of phreatic evaporation and zones with insignificant influence of phreatic evaporation, which has solved the problem of spatial variability of groundwater depth. As for the future scenario simulation, Figure 4 shows the annual temperature, precipitation, reference crop evapotranspiration, and groundwater depth variation during spring wheat growth period in the typical year. When the annual mean groundwater depth is less than 2.5 m, the groundwater depth during the spring wheat growth period is 1.29–2.61 m. However, when the annual mean groundwater depth is more than 2.5 m, the groundwater depth during the growth period is between 2.59 and 3.63 m. In practical application, the groundwater depth of 2.5 m in the previous year can be used to provide dynamic division so as to ensure that the irrigation schedule optimization can be better applied to shallow groundwater areas.

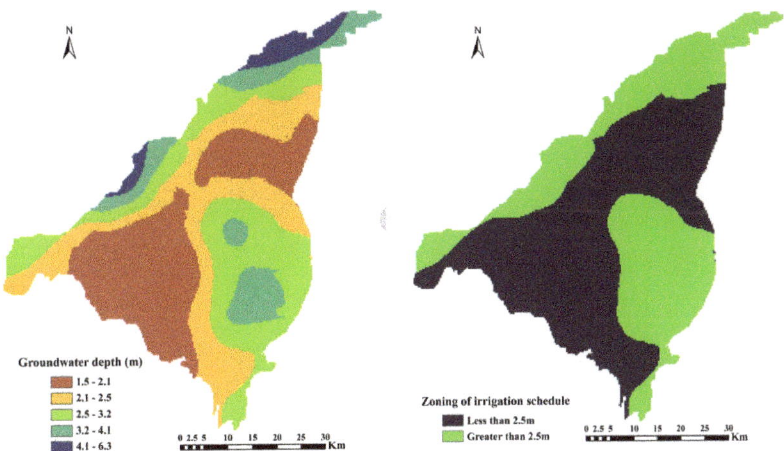

Figure 3. Spatial variation pattern of groundwater depth and irrigation schedule zoning in Jiefangzha region for the future scenario.

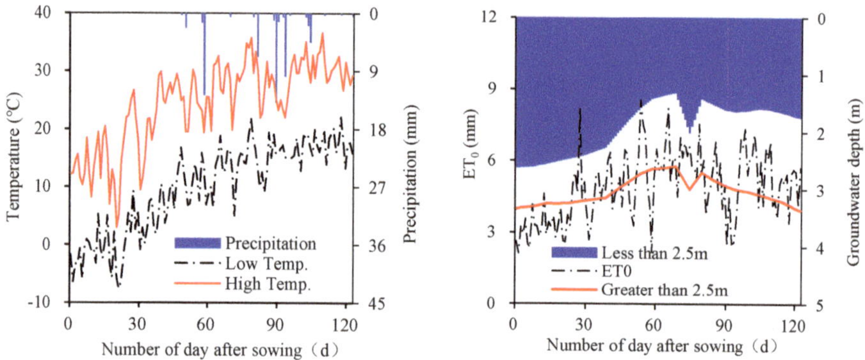

Figure 4. Annual temperature, precipitation, ET_0, and groundwater depth variation of the typical year for the future scenario.

2.6.3. Irrigation Schedule Scenarios

Because the study area is a canal irrigation area, there are only four times of irrigation in the growth period of spring wheat. According to the actual water distribution in the irrigation area, a total of irrigation scenarios was considered as rain-fed, one round of irrigation, two rounds of irrigation, three rounds of irrigation, and four rounds of irrigation for the four growth stages of spring wheat. When the total irrigation times of the whole growth period were determined, all possibilities for irrigation growth period were considered. There were 16 irrigation schedules, as shown in Table 4.

Table 4. Irrigation schedule scenarios.

Treatment	Tillering–Shooting	Shooting–Heading	Heading–Filling	Filling–Ripening	Irrigation Quota (mm)
T00					-
T11	√				60
T12		√			100
T13			√		100
T14				√	100
T21	√	√			160
T22	√		√		160
T23	√			√	160
T24		√	√		200
T25		√		√	200
T26			√	√	200
T31	√	√	√		260
T32	√	√		√	260
T33	√		√	√	260
T34		√	√	√	300
T44	√	√	√	√	360

Note: "√" means irrigation at this growing stage.

3. Results

3.1. Model Verification

It can be seen from Figure 5 and Table 5 that in calibration of the model, except for the T2–T4 treatments with slightly larger simulation values for the ripening stage, the simulated values for other growth stages are in good agreement with the measured soil moisture contents. The RMSE and the MAE between the simulated and measured soil moisture contents were less than 1.740% and 1.526%, respectively, and the R^2 was greater than 0.764, and the EF was greater than 0.722. For all irrigation treatments, in calibration of the model, the RMSE, MAE, R^2, and EF were 1.203%, 0.780%, 0.860, and 0.849, respectively. In model verification, the model simulation values satisfactorily reflected the change process of the measured soil moisture contents. As shown in Figure 6 and Table 5, the RMSE and MAE between simulated and measured values of soil moisture contents were below 1.802% and 1.429%, respectively, and the corresponding R^2 exceeded 0.651 and the EF was greater than 0.349. In model verification of all water treatments, the RMSE, MAE, R^2, and EF were 1.612%, 1.333%, 0.761, and 0.538, respectively. It can be seen that the fitting degree and accuracy of the soil moisture after model verification were both high, quite able to meet the simulation accuracy requirements of spring wheat soil water balance.

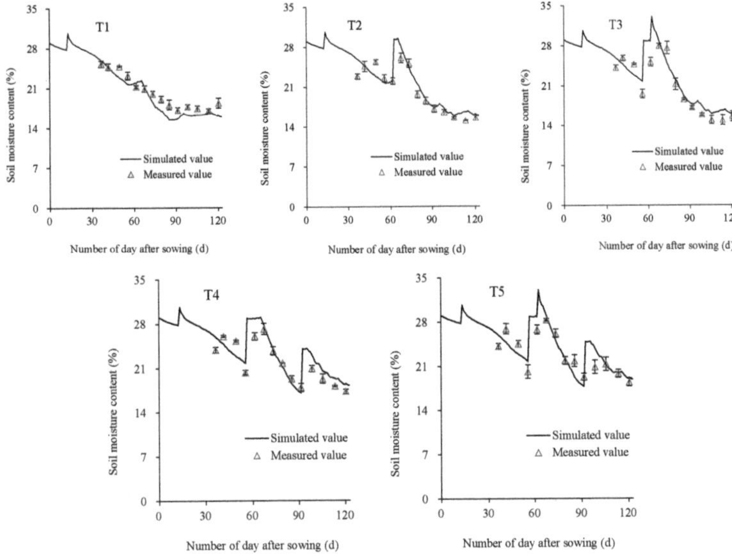

Figure 5. Simulated vs. measured values of spring wheat soil moisture content in model calibration.

Table 5. Evaluation indices of spring wheat soil moisture content simulation. RMSA: root mean square error.

		R^2	RMSE (%)	MAE (%)	MBE (%)	EF
	T1	0.927	1.481	1.290	−1.045	0.747
	T2	0.887	1.417	1.083	0.495	0.869
Model	T3	0.887	1.740	1.526	0.735	0.862
calibration	T4	0.825	1.522	1.343	0.473	0.784
	T5	0.764	1.635	1.383	0.066	0.722
	All treatments	0.860	1.203	0.780	0.037	0.849
	T1	0.710	1.578	1.278	−0.136	0.464
	T2	0.810	1.647	1.431	0.825	0.601
Model	T3	0.805	1.802	1.429	0.691	0.349
verification	T4	0.651	1.564	1.288	0.090	0.472
	T5	0.755	1.445	1.241	0.633	0.581
	All treatments	0.761	1.612	1.333	0.421	0.538

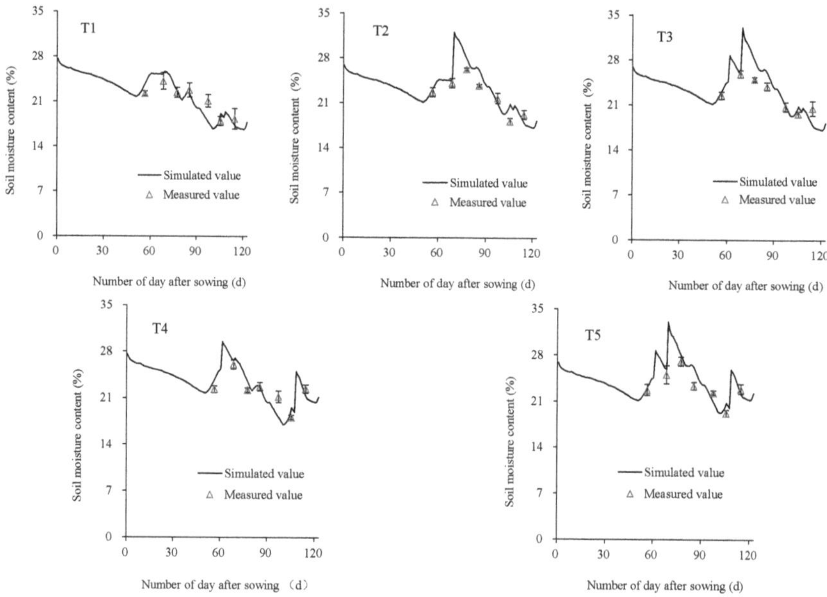

Figure 6. Simulated vs. measured values of spring wheat soil moisture content in model verification.

As can be seen from Figures 7 and 8, and Table 6, the simulated yields agreed well with the measured values. In model calibration, the RMSE, MAE, and MBE between the simulated and the observed values were 275.883 kg/hm^2, 246.190 kg/hm^2, −159.370 kg/hm^2 respectively, and the R^2 and EF were 0.985 and 0.976 respectively. In model verification, the RMSE, MAE, and MBE between the simulated and observed yields were 375.097 kg/hm^2, 242.402 kg/hm^2, and 145.004 kg/hm^2 respectively, and the R^2 and EF are 0.970 and 0.618 respectively. It can be seen that the RMSE and MAE between the simulated and observed values were less than 376 kg/hm^2 and 247 kg/hm^2, respectively, and the R^2 and EF were greater than 0.96 and 0.61 respectively. Hence, the model after verification is able to simulate satisfactorily spring wheat yield.

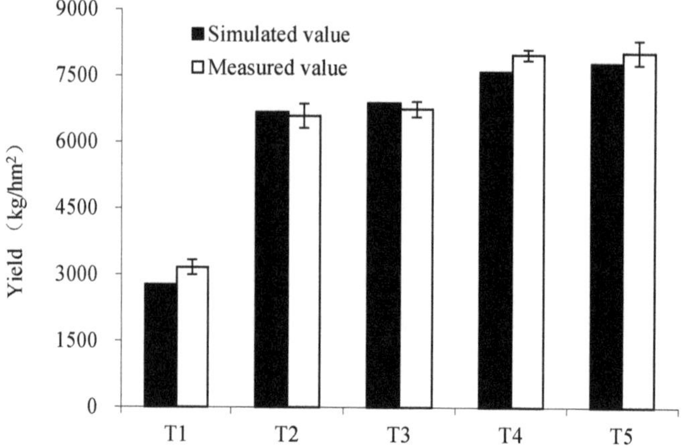

Figure 7. Simulated vs. measured values of spring wheat yield in model calibration.

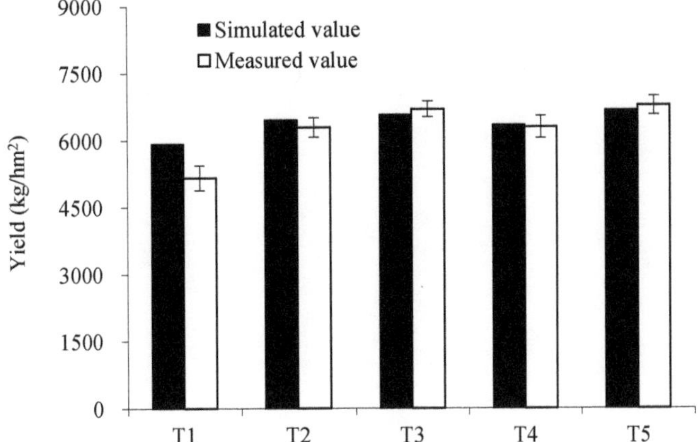

Figure 8. Simulated vs. measured values of spring wheat yield in model verification.

Table 6. Evaluation indices of spring wheat yield simulation.

	R^2	RMSE (kg/hm^2)	MAE (kg/hm^2)	MBE (kg/hm^2)	EF
Model calibration	0.985	275.883	246.190	−159.370	0.976
Model verification	0.970	357.097	242.402	145.004	0.618

In summary, the verified Aquacrop model is able to simulate the dynamic process of the soil moisture contents during the spring wheat growth period as well as the yield in shallow groundwater zones under different irrigation schedules. The model is useful in studying the relation between soil moisture contents and yield of spring wheat in shallow groundwater areas.

3.2. Water Consumption by Spring Wheat in Different Zones under Different Irrigation Schedules

Water consumption by spring wheat in different zones under different irrigation schedules is shown in Figure 9. As can be seen, where the groundwater depth was within 2.5 m, water consumption by rain-fed was 260 mm, and that by one round of irrigation was in the range of 284–387 mm. Water consumption by two rounds of irrigation was in the range of 326–424 mm. For three rounds and four rounds, the figures were 398–436 mm and 449 mm respectively. Where the groundwater depth was over 2.5 m, the water consumption by rain-fed was 210 mm. Water consumption by one round of irrigation was in the range of 234–326 mm. For two rounds and three rounds, the figures were in the range of 256–389 mm and 338–432 mm respectively. Water consumption by four rounds was 445 mm. It can be seen that within the irrigation quota of 360 mm, the water consumption of spring wheat increased with the irrigation quota. For a given irrigation number and a given irrigation quota, the water consumption varied greatly with the irrigation date. For the same irrigation schedule, less water was consumed when the groundwater depth exceeded 2.5 m than when the groundwater depth was less than 2.5 m, but the difference dwindled with the increase of irrigation quota.

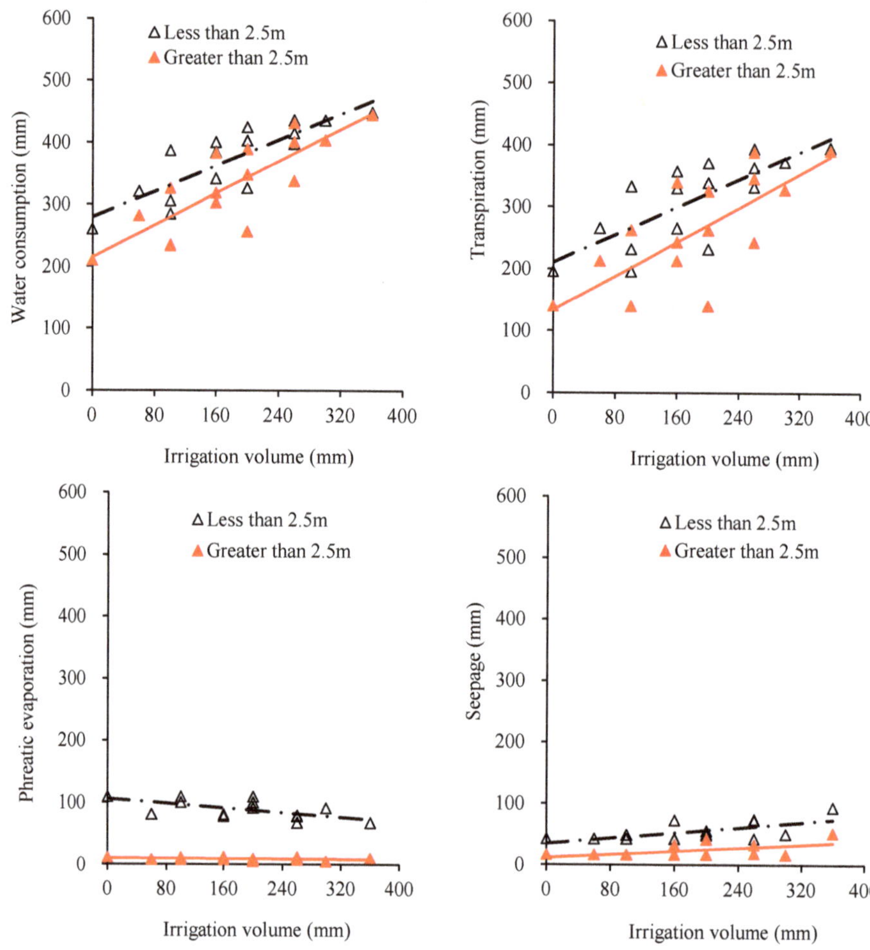

Figure 9. Water consumption by spring wheat in different zones under different irrigation schedules.

Where the groundwater depth was less than 2.5 m, the transpiration of rain-fed spring wheat was 196 mm, this figure was in the range of 196–333 mm for one round of irrigation or 232–370 mm for two rounds of irrigation, and for three rounds and four rounds of irrigation the transpiration was 332–394 mm and 396 mm respectively. Where the groundwater depth was greater than 2.5 m, the transpiration of rain-fed spring wheat was 140 mm, and for one round, two rounds, and three rounds of irrigation the figure was in the range of 140–263 mm, 140–339 mm, and 244–389 mm respectively. The transpiration was 391 mm for four rounds of irrigation. It can be seen that the way the transpiration of spring wheat varied with the groundwater depth parallels the relation between water consumption and the groundwater depth. It therefore follows that the change of transpiration is one of the most critical factors affecting the change of water consumption.

Where the groundwater depth was less than 2.5 m, the phreatic evaporation of rain-fed spring wheat was 109 mm, this figure was in the range of 81–109 mm for one round of irrigation or 77–109 mm for two rounds of irrigation, and for three rounds and four rounds of irrigation the phreatic evaporation was 67–91 mm and 67 mm respectively. Where the groundwater depth was greater than 2.5 m, the phreatic evaporation of rain-fed spring wheat was 12 mm, and for one round, two rounds, and three rounds of irrigation the figure was in the range of 8–12 mm, 5–12 mm, and 5–12 mm respectively. The phreatic evaporation was 10 mm for four rounds of irrigation. It could be seen that when the groundwater depth was more than 2.5 m, the phreatic evaporation of spring wheat was less than 12 mm, and it did not change much with the irrigation quota. When the groundwater depth was less than 2.5 m, the groundwater utilization decreased with the increase of irrigation quota.

Where the groundwater depth was less than 2.5 m, the seepage, in the case of rain-fed spring wheat, was 43 mm, this figure was in the range of 43–49 mm for one round of irrigation or 43–73 mm for two rounds of irrigation, and for three rounds and four rounds of irrigation the seepage was 43–74 mm and 94 mm respectively. Where the groundwater depth was greater than 2.5 m, the seepage, in the case of rain-fed spring wheat, was 17 mm, and for one round, two rounds, and three rounds of irrigation this figure was 17 mm, 17–42 mm, and 17–33 mm respectively. The seepage was 53 mm for four rounds of irrigation. It could be seen that, within the net irrigation quota of 360 mm, the amount of seepage increased with the irrigation quota; with the same irrigation schedule, when the groundwater depth was more than 2.5 m, the seepage was smaller than when the depth was less than 2.5 m.

3.3. Yield of Spring Wheat in Different Zones under Different Irrigation Schedules

The yields of spring wheat in different zones under different irrigation schedules are shown in Figure 10. Where the groundwater depth was less than 2.5 m, the yield of rain-fed spring wheat was 2505 kg/hm^2, and this figure was in the range of 2505–6283 kg/hm^2, for one round of irrigation, 4384–7091 kg/hm^2 for two rounds of irrigation, or 6272–7640 kg/hm^2 for three rounds of irrigation. For four rounds of irrigation, the yield was 7672 kg/hm^2. Where the groundwater depth was greater than 2.5 m, there was zero yield of the rain-fed spring wheat. The yield of spring wheat for one round of irrigation was in the range of 0–4844 kg/hm^2, and this figure was in the range of 0–6498 kg/hm^2 for two rounds of irrigation or 4548–7600 kg/hm^2 for three rounds of irrigation. For four rounds of irrigation, the yield of spring wheat was 7650 kg/hm^2. Where the groundwater depth was less than 2.5 m, the yield of T12 for one round of irrigation was up to 6283 kg/hm^2, the yield of T24 for two rounds of irrigation was up to 7091 kg/hm^2, and the yield of T31 for three rounds of irrigation was up to 7640 kg/hm^2. Where the groundwater depth was more than 2.5 m, the yield of T12 for one round of irrigation was up to 4844 kg/hm^2, the yield of T21 for two rounds of irrigation was up to 6498 kg/hm^2, and the yield of T3 for three rounds of irrigation was up to 7600 kg/hm^2. It could be seen that with the increase of irrigation quota, the yield of spring wheat generally increased. The timing of irrigation was especially important if the total times of irrigation remained constant. As shallow groundwater replenished available water to the crop, the yield in shallow groundwater depth zones was higher than that in deeper groundwater depth zones under the same irrigation schedule. In light of this, in the case of one round of irrigation, it is important to meet the wheat water demand at shooting–heading stage. Where the groundwater depth is less than 2.5 m, in order to take greater advantage of groundwater, the key is to satisfy water demand at the shooting–heading and heading–filling stages in the case of two rounds of irrigation, and where the groundwater depth is more than 2.5 m, it is important to satisfy the wheat water demand at the tillering–shooting and shooting–heading stages. In the case of three rounds of irrigation, the key is to satisfy the water demand at the tillering–shooting, shooting–heading, and heading–filling stages.

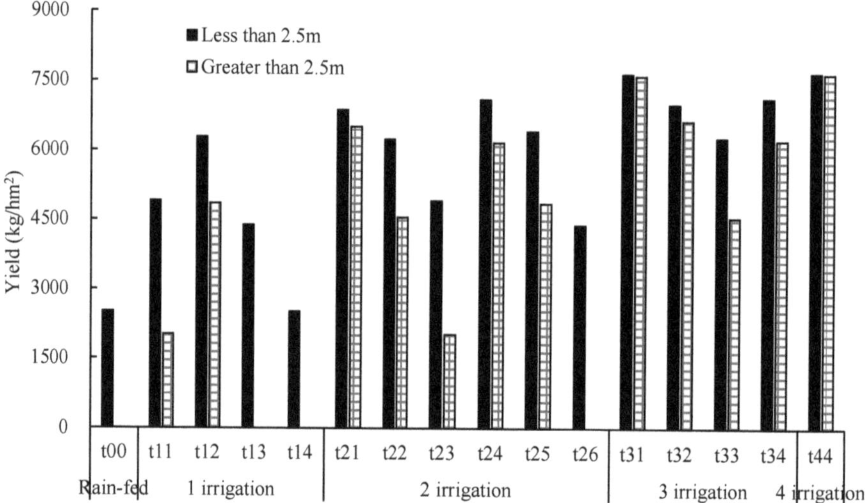

Figure 10. Yields of spring wheat in different zones under different irrigation schedules.

3.4. Optimization of Spring Wheat Irrigation Schedule Considering Groundwater Spatial Variability

The sensitivity indices and test parameters of the spring wheat water production function model are shown in Table 7. When the groundwater depth was greater than 2.5 m, the absolute values of the sensitivity indices, evaluated by Jensen and Minhas models, at some growth stages were greater than 1, in conflict with the theoretical value. Therefore, the two models are not suitable for simulating the relationship between the yield and water consumption at the growth stages when the groundwater depth is greater than 2.5 m. In the three models of Blank, Stewart, and Singh, the Stewart model gave the largest R^2, which was up to 0.98, and the lowest RMSE, which was only 410.58 kg·hm^{-2}. Therefore, it is advisable to take Stewart model as the water production function of spring wheat at the growth stages when the groundwater depth is greater than 2.5 m. From the results given by the Stewart model, the sensitivity coefficient for the tillering–shooting stage was up to 0.7614 when the groundwater depth was greater than 2.5 m, suggesting that it is most sensitive to water shortage at this stage. The sensitivity coefficient was 0.6691 for the shooting–heading stage or 0.5060 for the heading–filling stage. The minimum sensitivity coefficient was −0.0109, which was for the filling–ripening stage, indicating that it is not sensitive to water shortage at this growth stage. When the groundwater depth was less than 2.5 m, the values of the sensitivity indices, evaluated by Minhas and Steward models, at some growth stages were greater than 1, in conflict with the theoretical value. Therefore, the two models are not suitable for simulating the relationship between the yield and the water consumption at the growth stages when the groundwater depth is less than 2.5 m. In the three models of Jensen, Blank, and Singh, the Jensen model gave the largest R^2, which was up to 0.99, and the lowest RMSE, which was only 165.32 kg·hm^{-2}. Therefore, it is advisable to take the Jensen model as the water production function of spring wheat at the growth stages when the groundwater depth is less than 2.5 m. From the results of the Jensen model, when the groundwater depth was less than 2.5 m the sensitivity index was up to 0.9930 for the tillering–shooting stage, was 0.6202 for the heading–filling stage, but was only 0.3591 for the shooting–heading stage. The sensitivity index for the filling–ripening stage was negative, indicating that this stage, too, is not sensitive to water shortage. By comparing the sensitivity for different spring wheat growth stages under different zones, we can see a big difference between the two zones at the shooting–heading stage. When the groundwater depth is greater than 2.5 m, spring wheat is more sensitive to water shortage, while when the depth is less than 2.5 m, the sensitivity to water shortage at this stage is lower because the groundwater supplies the crops with available water. Therefore, the challenge of greater spatial variation of groundwater can be practically taken care of by zoning method. Spring wheat is most sensitive to water deficiency at the tillering–shooting stage, is less sensitive to water deficiency at the heading–filling stage, and is least sensitive to water deficiency at the filling–ripening stage, irrespective of the zone. It can be seen that the water-sensitive results at different growth stages under different zones suggest an agreement with the above-described order of importance of satisfying water demand at different growth stages.

Table 7. Sensitivity indices and test parameters of water production function model of spring wheat at different growth stages.

Groundwater Depth	Model	Tillering–Shooting	Shooting–Heading	Heading–Filling	Filling–Ripening	R^2	RMSE (kg/hm^2)	MAE (kg/hm^2)	MBE (kg/hm^2)
Greater than 2.5 m	Jensen	−2.4901	4.3914	−0.4899	0.0252	0.77	1565.50	1059.55	128.87
	Minhas	38.0809	4.9308	0.3486	0.0112	0.94	692.37	441.15	141.82
	Blank	−0.3555	0.8647	0.4171	0.0344	0.97	458.34	327.79	−4.24
	Stewart	0.7614	0.6691	0.506	−0.0109	0.98	410.58	306.51	22.58
	Singh	−0.6875	0.8138	0.6344	0.0714	0.96	530.17	459.80	0.11
Less than 2.5 m	Jensen	0.9930	0.3591	0.6202	−0.0280	0.99	165.32	136.03	30.87
	Minhas	37.7041	−0.0287	0.8584	0.0743	0.95	487.17	319.54	−265.34
	Blank	−0.0437	0.4508	0.5864	−0.0434	0.98	198.21	166.18	−2.04
	Stewart	1.0475	0.3265	0.5562	−0.0134	0.99	145.73	125.48	−2.28
	Singh	−0.7359	0.9530	0.5868	0.0490	0.98	231.57	169.90	−0.54

With the verified Aquacrop as the technical support and the soil moisture content of the root layer as the control index, lower irrigation limits were set in light of the sensitivity variation across the growth stages under different groundwater depths conditions. Where the groundwater depth was greater than 2.5 m, no irrigation was given at the sowing–tillering and filling–ripening stags, but irrigation started when the soil moisture of root layer dropped below the lower irrigation limit at the tillering–shooting stage or when the content dropped below 10% of this lower irrigation limit at the shooting-filling. Where the groundwater depth was less than 2.5 m, no irrigation was given at the sowing–tillering stage and the filling–ripening stage, but irrigation started once the soil moisture of root layer dropped below the lower irrigation limit at the tillering–shooting stage or when it dropped below 20% of this lower irrigation limit at the shooting-filling stage. With per irrigation quota of 60–120 mm, the optimized irrigation schedules under different groundwater depth conditions were developed. Where the groundwater depth was greater than 2.5 m, there were two rounds of irrigation both at the tillering–shooting stage and the shooting–heading stage, with the irrigation quota being 300 mm, the water consumption being 486 mm, the yield being 8236 kg/hm^2, and the water productivity being 1.694 kg/m^3. Where the groundwater depth was less than 2.5 m, there were two rounds of irrigation at the tillering–shooting stage and one round of irrigation at the shooting–heading stage, with the irrigation quota of 240 mm, the water consumption of 474 mm, the yield of 8014 kg/hm^2, and the water productivity of 1.690 kg/m^3. Still, throughout the growth stages of spring wheat, full irrigation schedules were developed for spring wheat under different groundwater depth conditions such that irrigation started once the soil moisture content of the root layer dropped below the lower irrigation limit with the per irrigation quota being 60–120 mm. Where the groundwater depth was greater than 2.5 m, the irrigation quota was 360 mm and the water consumption was 492 mm, with the yield of 8343 kg/hm^2 and the water productivity of 1.697 kg/m^3. Where the groundwater depth was less than 2.5 m, the irrigation quota was 320 mm and the water consumption was 493 mm, with the yield of 8384 kg/hm^2 and the water productivity of 1.701 kg/m^3.

4. Discussion

4.1. Water Saving Performance Analysis

The soil water balance of spring wheat in different zones under different irrigation schedules is shown in Table 8. Where that the groundwater depth was less than 2.5 m, the irrigation quota, water consumption, phreatic evaporation, seepage, and soil water utilization under the current irrigation schedule were 335 mm, 440 mm, 65 mm, 127 mm, and 65 mm respectively. Under the full irrigation schedule these figures were 320 mm, 493 mm, 55 mm, 70 mm, and 85 mm respectively. Under the optimized irrigation schedule, they were 240 mm, 474 mm, 62 mm, 68 mm, and 139 mm respectively. It could be seen that under the optimized irrigation schedule, the seepage dropped by 46% and the soil water use increased by 114%. Where the groundwater depth was more than 2.5 m, the irrigation quota, water consumption, phreatic evaporation, seepage, and soil water utilization under the current irrigation schedule were 335 mm, 438 mm, 8 mm, 87 mm, and 102 mm respectively. Under the full irrigation schedule these figures were 360 mm, 492 mm, 7 mm, 84 mm, and 129 mm respectively. Under the optimized irrigation schedule, they were 300 mm, 486 mm, 7 mm, 47 mm, and 147 mm respectively. It could be seen that under the optimized irrigation schedule, the seepage dropped by 46% and the soil water use increased by 44%. The spring wheat soil water balance table shows that the optimized irrigation schedule cut the seepage loss and improved the soil water use on the current irrigation schedule. As far as the groundwater depth is concerned, a shallow depth has a more significant effect on reducing seepage and increasing soil water use. Where the groundwater depth was less than 2.5 m, the current irrigation schedule had a slightly higher irrigation quota than the full irrigation schedule; where the groundwater depth was greater than 2.5 m, the current irrigation schedule had a slightly lower irrigation quota than the full irrigation schedule. This suggests that the regional agricultural water saving potential lies mainly in the optimization of the crop irrigation schedule.

Table 8. Soil water balance of spring wheat in different zones under different irrigation schedules.

Groundwater Depth	Irrigation Schedule	Input (mm)		Output (mm)			Soil Water Use (mm)
		Precipitation	Irrigation Quota	Phreatic Evaporation	Seepage	Water Consumption	
Greater than 2.5 m	Current irrigation schedule	57	335	8	87	438	102
	Full irrigation	57	360	7	84	492	129
	Optimized irrigation schedule	57	300	7	47	486	147
Less than 2.5 m	Current irrigation schedule	57	335	65	127	440	65
	Full irrigation	57	320	55	70	493	85
	Optimized irrigation schedule	57	240	62	68	474	139

The water saving performance of the optimized irrigation schedule for spring wheat in each zone is shown in Figure 11. As can be seen, compared with the current irrigation schedule, when the groundwater depth was less than 2.5 m, with the optimized irrigation schedule the irrigation water consumption dropped by 95 mm, the yield increased by 377 kg/hm^2, the water consumption grew by

35 mm, the transpiration was up by 48 mm, the water productivity fell by 0.04 kg/m^3, the irrigation water productivity was higher by 0.31 kg/m^3, while with the full irrigation schedule the irrigation water quota dropped by 15 mm, the yield increased by 747 kg/hm^2, the water consumption grew by 53 mm, the transpiration was up by 67 mm, and the water productivity and the irrigation water productivity remained unchanged. When the groundwater depth was more than 2.5 m, with the optimized irrigation schedule the irrigation water quota dropped by 35 mm, the yield increased by 581 kg/hm^2, the water consumption grew by 49 mm, the transpiration was up by 63 mm, the water productivity fell by 0.06 kg/m^3, the irrigation water productivity grew by 0.46 kg/m^3, while with the full irrigation schedule the irrigation water quota dropped by 25 mm, the yield increased by 729 kg/hm^2, the water consumption grew by 54 mm, the transpiration was up by 68 mm, the water productivity fell by 0.04 kg/m^3, and the irrigation water productivity grew by 0.04 kg/m^3. It could be seen that where the groundwater depth was less than 2.5 m, with the optimized irrigation schedule the irrigation quota of spring wheat fell by 28%, the yield increased by 5%, the irrigation water productivity grew by 20%, the additional water consumption was all used for crop transpiration, and the water productivity reduction was less than 3%; where the groundwater depth was greater than 2.5 m, with the optimized irrigation schedule the irrigation quota was reduced by 10%, the yield increased by 8%, the irrigation water productivity grew by 20%, the additional water consumption was all used for crop transpiration, and the water productivity reduction was less than 4%. From the soil water balance data, the optimized irrigation schedule's water-saving effect is mainly seen in greater yield and higher irrigation water productivity, lower irrigation quota, less ineffective seepage and soil evaporation, and substantial increase in soil water use, while the amount of additional water consumption was all used for crop transpiration. Compared with the groundwater depth over 2.5 m, when this depth was less than 2.5 m, with the optimized irrigation schedule the irrigation quota dropped by 20% and the soil water use was significantly improved.

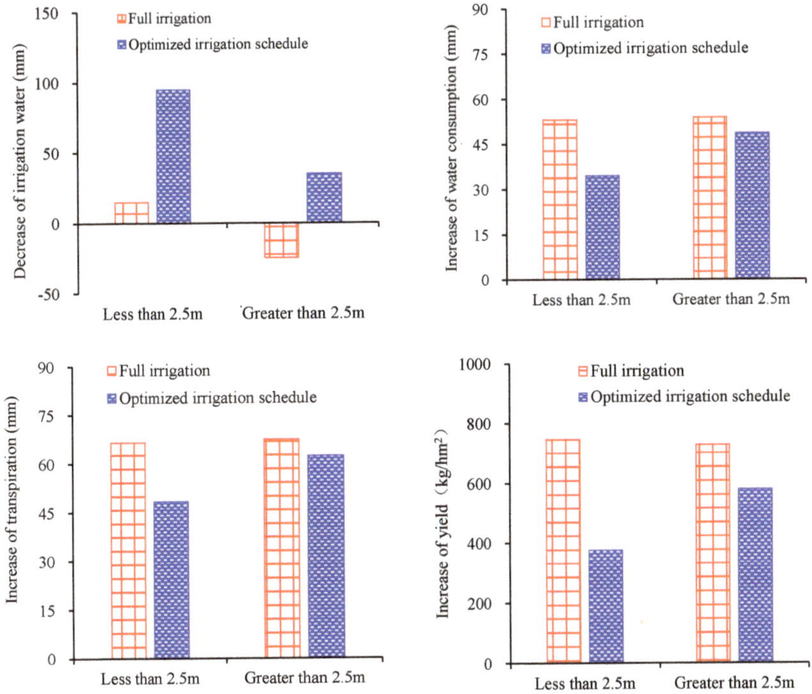

Figure 11. *Cont.*

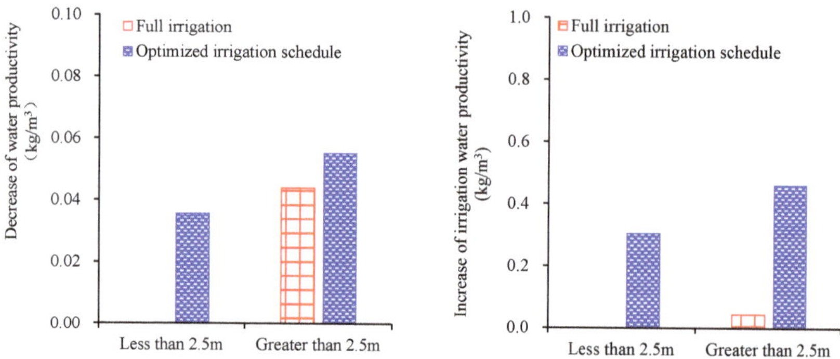

Figure 11. Water saving performance of optimized irrigation schedule for spring wheat in each zone.

4.2. Water Consumption Characteristics

Figure 12 shows the daily water consumption and cumulative water consumption of spring wheat at each growth stage in each zone under the optimized irrigation schedule. It could be seen that the average daily water consumption increased rapidly from 1.87–1.89 mm at the sowing–tillering stage to a maximum of 5.82–5.89 mm at the tillering–heading stage, and then gradually fell to 4.95–5.16 mm at the heading–filling stage and further to 2.66–3.08 mm at the filling–ripening stage. The difference in daily water consumption at different growth stages did not vary much between the two groundwater depths. When the groundwater depth was less than 2.5 m, the cumulative water consumption at the sowing–tillering, tillering–shooting, shooting–heading, heading–filling, and filling–ripening stages was 80 mm, 220 mm, 349 mm, 424 mm, and 474 mm respectively. When the groundwater depth was greater than 2.5 m, the cumulative water consumption at these stages was 81 mm, 221 mm, 350 mm, 428 mm, and 486 mm respectively. It could be seen that, irrespective of the groundwater depths, the water consumption characteristics of the optimized irrigation schedule were basically the same—the average daily water consumption peaked at the tillering–heading stage, with a combined water consumption accounting for 55.4%–56.7% of the total water consumption during the whole growth period.

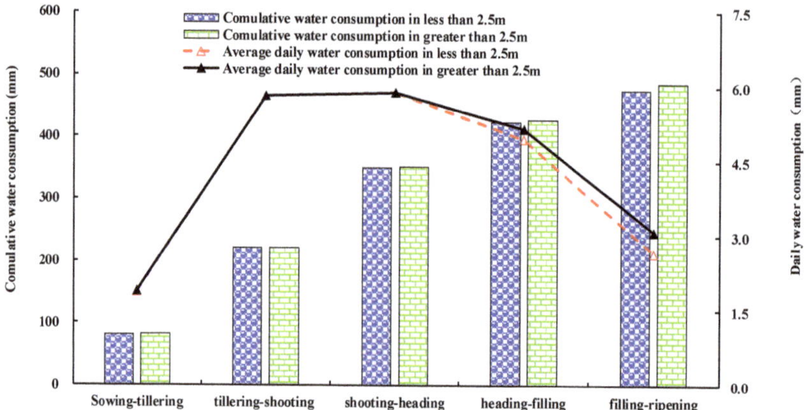

Figure 12. Spring wheat water consumption characteristics under optimized irrigation schedule.

5. Conclusions

(1) The Aquacrop model after verification can satisfactorily simulate the dynamic process of the soil moisture content during the growth period of spring wheat and its yield under different

irrigation schedules respectively, which can be used to investigate the water-yield response mechanism of spring wheat.

(2) The average groundwater table during the spring wheat growing season makes a critical precondition for the simulation of the zoned irrigation schedule, which gives the regional representativeness and feasibility for the optimized irrigation schedules, and importantly provides a solution to the disruption between the spatial variability and the optimization of the irrigation schedule in shallow groundwater areas.

(3) For an irrigation quota within 360 mm, as the irrigation quota increases, the water consumption, seepage, and yield all increase, while the groundwater utilization presents a decreasing trend. In order to get the greater yields, the choice of irrigation schedule is especially important.

(4) Where the groundwater depth is greater than 2.5 m, two rounds of irrigation are made at both the tillering–shooting stage and the shooting–heading stage. Where the groundwater depth is less than 2.5 m, two rounds of irrigation are made at the tillering–shooting stage and one round of irrigation is made at the shooting–heading stage.

(5) The main water-saving effect of the optimized irrigation schedule is that the spring wheat yield, the soil moisture availability, and the irrigation water productivity increase while, the irrigation amount and the ineffective seepage decrease, from which the additional water consumption can be fully used for crop transpiration, being a kind of effective water consumption.

Author Contributions: All authors read and approved the manuscript. Z.P. put up with the main idea of this research. conceptualization, Z.P.; methodology, Z.P., B.Z., Y.L. and Z.W.; software, Z.P.; validation, Z.P., B.Z., and J.C.; formal analysis, Z.P. and H.C.; writing—original draft preparation, Z.P.; writing—review and editing, Z.P., B.Z., Z.W., and H.C.; project administration, Z.P. and Y.L.; funding acquisition, B.Z.

Funding: This study was supported by the National Key R&D Program of China (2018YFC0407703), the IWHR Research & Development Support Program (ID0145B082017, ID0145B742017, ID0145B102019 and ID01881910), The Key R&D projects of Ningxia Hui Autonomous Region (2018BBF02022), the Chinese National Natural Science Fund (51822907 and 51979287), the Special Fund of State Key Laboratory of Simulation and Regulation of Water Cycle in River Basin, and China Institute of Water Resources and Hydropower Research (SKL2018CG03).

Acknowledgments: The authors greatly appreciate the anonymous reviewers and the academic editor for their careful comments and valuable suggestions to improve the manuscript.

Conflicts of Interest: The authors declare no conflict of interest.

References

1. Wang, P.; Yu, J.J.; Pozdniakov, S.P.; Grinevsky, S.O. Shallow groundwater dynamics and its driving forces in extremely arid areas: A case study of the lower Heihe River in northwestern China. *Hydrol. Process.* **2014**, *28*, 1539–1553. [CrossRef]

2. Satchithanantham, S.; Krahn, V.; Ranjan, R.S.; Sager, S. Shallow groundwater uptake and irrigation water redistribution within the potato root zone. *Agric. Water Manag.* **2014**, *132*, 101–110. [CrossRef]

3. Ghamarnia, H.; Golamian, M.; Sepehri, S.; Arji, I. Shallow groundwater use by Safflower (*Carthamus tinctorius* L.) in a semi-arid region. *Irrig. Sci.* **2011**, *29*, 147–156. [CrossRef]

4. Gao, X.Y.; Huo, Z.L.; Qu, Z.Y. Modeling contribution of shallow groundwater to evapotranspiration and yield of maize in an arid area. *Sci. Rep.* **2017**, *7*, 1–13. [CrossRef] [PubMed]

5. Kong, F.R.; Qu, Z.Y.; Liu, Y.J.; Ma, Y.X. Experimental research on the effect of different kinds of groundwater buried depth on soil water, salinity and crop growth. *China Rural Water Hydropower* **2009**, *5*, 44–48. (In Chinese)

6. Kruse, E.G.; Champion, D.F.; Cuevas, D.L.; Yoder, R.E.; Young, D. Crop water use from shallow, saline water tables. *Trans. Asae* **1993**, *36*, 697–707. [CrossRef]

7. Wang, X.H.; Hou, H.B. The influence of shallow groundwater on crops growth laws. *J. Irrig. Drain.* **2006**, *25*, 13–16. (In Chinese)

8. Zhang, C.X. Study on groundwater utilization by dry crops. *J. Irrig. Drain.* **1989**, *8*, 20–24. (In Chinese)

9. Yang, P.Y.; Wu, B.; Wang, S.X.; Dong, X.G.; Liu, L. Research on irrigation schedule of cotton drip irrigation under plastic film based on the different ground water table in arid areas. *Agric. Res. Arid Areas* **2014**, *32*, 76–82. (In Chinese)

10. Zhuang, Y.; Lu, J.; Jiao, X. Effect of groundwater table on cotton irrigation regulation. *Water Sav. Irrig.* **2015**, *3*, 15–17. (In Chinese)

11. Wang, B.L. Experimental study on the utilization of shallow groundwater for spring wheat. *Agric. Sci. Technol.* **2011**, *12*, 108–132.

12. Liu, Z.D.; Liu, Z.G.; Yu, J.H.; Nan, J.Q.; Qin, A.Z.; Xiao, J.F. Effects of groundwater depth on maize growth and water use efficiency. *J. Drain. Irrig. Mach. Eng.* **2014**, *32*, 617–624. (In Chinese)

13. Xue, J.Y.; Guan, H.; Huo, Z.L.; Wang, F.X.; Huang, G.H.; Boll, J. Water saving practices enhance regional efficiency of water consumption and water productivity in an arid agricultural area with shallow groundwater. *Agric. Water Manag.* **2017**, *194*, 78–89. [CrossRef]

14. Ghamarnia, H.; Farmanifard, M. Yield production and water-use efficiency of wheat (*Triticum aestivum* L.) cultivars under shallow groundwater use in semi-arid region. *Arch. Agron. Soil Sci.* **2014**, *60*, 1677–1700. [CrossRef]

15. Ren, D.Y.; Xu, X.; Huang, Q.Z.; Huo, Z.L. Analyzing the Role of Shallow Groundwater Systems in the Water Use of Different Land-Use Types in Arid Irrigated Regions. *Water* **2018**, *10*, 634. [CrossRef]

16. Wang, X.W.; Huo, Z.L.; Guan, H.; Guo, P. Drip irrigation enhances shallow groundwater contribution to crop water consumption in an arid area. *Hydrol. Process.* **2018**, *32*, 747–758. [CrossRef]

17. Chen, H.; Liu, Z.Y.; Huo, Z.L.; Qu, Z.Y.; Xia, Y.H.; Fernald, A. Impacts of agricultural water saving practice on regional groundwater and water consumption in an arid region with shallow groundwater. *Environ. Earth Sci.* **2016**, *75*, 1204. [CrossRef]

18. Ebrahimi, H.; Ghazavi, R.; Karimi, H. Estimation of Groundwater Recharge from the Rainfall and Irrigation in an Arid Environment Using Inverse Modeling Approach and RS. *Water Resour. Manag.* **2016**, *30*, 1939–1951. [CrossRef]

19. Gao, X.Y.; Bai, Y.N.; Huo, Z.L.; Xu, X.; Huang, G.H.; Xia, Y.H.; Steenhuisc, T.S. Deficit irrigation enhances contribution of shallow groundwater to crop water consumption in arid area. *Agric. Water Manag.* **2017**, *185*, 116–125. [CrossRef]

20. Xue, J.Y.; Huo, Z.L.; Wang, F.X.; Kang, S.Z.; Huang, G.H. Untangling the effects of shallow groundwater and deficit irrigation on irrigation water productivity in arid region: New conceptual model. *Sci. Total Environ.* **2018**, *619*, 1170–1182. [CrossRef]

21. Gao, X.Y.; Huo, Z.L.; Xu, X.; Qu, Z.Y. Shallow groundwater plays an important role in enhancing irrigation water productivity in an arid area: The perspective from a regional agricultural hydrology simulation. *Agric. Water Manag.* **2018**, *208*, 43–58. [CrossRef]

22. Karimov, A.K.; Šimůnek, J.; Hanjra, M.A.; Avliyakulov, M.; Forkutsa, I. Effects of the shallow water table on water use of winter wheat and ecosystem health: Implications for unlocking the potential of groundwater in the Fergana Valley (Central Asia). *Agric. Water Manag.* **2014**, *131*, 57–69. [CrossRef]

23. Liu, J.; Sun, S.K.; Wang, Y.B.; Zhao, X.N. Evaluation of crop production, trade, and consumption from the perspective of water resources: A case study of the Hetao irrigation district, China, for 1960–2010. *Sci. Total Environ.* **2015**, *505*, 1174–1181. [CrossRef] [PubMed]

24. Bai, L.L.; Cai, J.B.; Liu, Y.; Chen, H.; Zhang, B.Z.; Huang, L.X. Responses of field evapotranspiration to the changes of cropping pattern and groundwater depth in large irrigation district of Yellow River basin. *Agric. Water Manag.* **2017**, *166*, 17–32. [CrossRef]

25. Du, J.; Yang, P.L.; Ren, S.M.; Li, Y.K.; Wang, Y.Z.; Yuan, X.Q.; Li, X.Y.; Du, J. The canal-lining project on groundwater and ecological environment in Hetao irrigation District of Inner Mongolia. *Chin. J. Appl. Ecol.* **2011**, *22*, 144–150. (In Chinese)

26. Li, Z.Z.; Fu, X.J.; Tian, Y.; Zhen, Q. Study on the Change of Soil Moisture and Groundwater by Jiefangzha Region in Hetao Irrigation District. *Inn. Mong. Water Resour.* **2018**, *5*, 16–17. (In Chinese)

27. Allen, R.; Periera, L.; Raes, D.; Smith, M. *Crop Evapotranspriation: Guidelines for Computing Crop Water Requirement*; FAO Irrigation and Drainage Paper No.56; FAO: Rome, Italy, 1998.

28. Hsiao, T.; Heng, L.; Steduto, P.; Rojas-Lara, B.; Raes, D.; Fereres, E. AquaCrop—The FAO Crop Model to Simulate Yield Response to Water: III. Parameterization and Testing for Maize. *Agron. J.* **2009**, *101*, 448–459. [CrossRef]

29. Raes, D.; Steduto, P.; Hsiao, T.; Fereres, E. AquaCropThe FAO Crop Model to Simulate Yield Response to Water: II. Main Algorithms and Software Description. *Agron. J.* **2009**, *101*, 438–447. [CrossRef]

30. Steduto, P.; Hsiao, T.; Raes, D.; Fereres, E. AquaCrop—The FAO Crop Model to Simulate Yield Response to Water: I. Concepts and Underlying Principles. *Agron. J.* **2009**, *101*, 426–437. [CrossRef]

31. Andarzian, B.; Bannayan, M.; Steduto, P.; Mazraeh, H.; Barati, M.E.; Barati, M.A.; Rahnama, A. validation and testing of the Aquacrop model under full and deficit irrigated wheat production in Iran. *Agric. Water Manag.* **2011**, *100*, 1–8. [CrossRef]

32. Kahimba, F.C.; Bullock, P.R.; Sri-Ranjan, R.; Cutforth, H.W. Evaluation of the SolarCalc model for simulating hourly and daily incoming solar radiation in the Northern Great Plains of Canada. *Can. Biosyst. Eng.* **2009**, *51*, 1–11.

33. Willmott, C.J. Some comments on the evaluation of model performance. *Bull. Am. Meteorol. Soc.* **1982**, *63*, 1309–1313. [CrossRef]

34. Willmott, C.J.; Maasuura, K. Advantages of the meanabsolute error (MAE) over the root mean square error (RMSE) in assessing average model performance. *Clim. Res.* **2005**, *30*, 79–82. [CrossRef]

35. Zhao, S.Z.; Kong, F.J.; Wang, X.K.; Li, S.B.; Meng, K.W. Confirming of critical depth of groundwater level and discussion on it's significance-take Hetao Irrigation Area for example. *J. Inn. Mong. Agric. Univ.* **2008**, *29*, 164–167. (In Chinese)

36. Ruan, B.Q.; Han, Y.P.; Jiang, R.F.; Xu, F.R. Appropriate water saving extent for ecological vulnerable area. *J. Hydraul. Eng.* **2008**, *39*, 809–814. (In Chinese)

 © 2019 by the authors. Licensee MDPI, Basel, Switzerland. This article is an open access article distributed under the terms and conditions of the Creative Commons Attribution (CC BY) license (http://creativecommons.org/licenses/by/4.0/).

 *water*

Article

Saturated Hydraulic Conductivity Estimation Using Artificial Neural Networks

Josué Trejo-Alonso [1], Carlos Fuentes [2,*], Carlos Chávez [1,*], Antonio Quevedo [2], Alfonso Gutierrez-Lopez [3] and Brandon González-Correa [4]

1. Water Research Center, Department of Irrigation and Drainage Engineering, Autonomous University of Queretaro, Cerro de las Campanas SN, Col. Las Campanas, Queretaro 76010, Mexico; josue.trejo@uaq.mx
2. Mexican Institute of Water Technology, Paseo Cuauhnáhuac Núm. 8532, Jiutepec 62550, Mexico; jose_quevedo@tlaloc.imta.mx
3. Water Research Center, Centro de Investigaciones del Agua-Queretaro (CIAQ), International Flood Initiative, Latin-American and the Caribbean Region (IFI-LAC), International Hydrological Programme (IHP-UNESCO), Universidad Autonoma de Queretaro, Queretaro 76010, Mexico; alfonso.gutierrez@uaq.mx
4. Engineering Faculty, Autonomous University of Queretaro, Cerro de las Campanas SN, Col. Las Campanas, Queretaro 70610, Mexico; gonzalezcorrea.fm1@gmail.com
* Correspondence: cbfuentesr@gmail.com (C.F.); chagcarlos@uaq.mx (C.C.); Tel.: +52-442-192-1200 (ext. 6036) (C.C.)

Abstract: In the present work, we construct several artificial neural networks (varying the input data) to calculate the saturated hydraulic conductivity (K_S) using a database with 900 measured samples obtained from the Irrigation District 023, in San Juan del Rio, Queretaro, Mexico. All of them were constructed using two hidden layers, a back-propagation algorithm for the learning process, and a logistic function as a nonlinear transfer function. In order to explore different arrays for neurons into hidden layers, we performed the bootstrap technique for each neural network and selected the one with the least Root Mean Square Error (RMSE) value. We also compared these results with pedotransfer functions and another neural networks from the literature. The results show that our artificial neural networks obtained from 0.0459 to 0.0413 in the RMSE measurement, and 0.9725 to 0.9780 for R^2, which are in good agreement with other works. We also found that reducing the amount of the input data offered us better results.

Keywords: modeling water flow; gravity irrigation; infiltration process; artificial intelligence

Citation: Trejo-Alonso, J.; Fuentes, C.; Chávez, C.; Quevedo, A.; Gutierrez-Lopez, A.; González-Correa, B. Saturated Hydraulic Conductivity Estimation Using Artificial Neural Networks. *Water* 2021, 13, 705. https://doi.org/10.3390/w13050705

Academic Editors: George Kargas, Petros Kerkides and Paraskevi Londra

Received: 10 February 2021
Accepted: 27 February 2021
Published: 5 March 2021

Publisher's Note: MDPI stays neutral with regard to jurisdictional claims in published maps and institutional affiliations.

Copyright: © 2020 by the authors. Licensee MDPI, Basel, Switzerland. This article is an open access article distributed under the terms and conditions of the Creative Commons Attribution (CC BY) license (https://creativecommons.org/licenses/by/4.0/).

1. Introduction

Soil water movement is important in several fields, like irrigation, drainage, hydrology, and agriculture [1]. Among all the measurable quantities in soils, one of the most important is the saturated hydraulic conductivity (K_S), defined as the ability to transmit water throughout the saturated zone [2,3], which is highly correlated with the optimization of the flow rate applied to the border or furrow in the gravity irrigation [2–6]. Although this property is measured easily in a laboratory, or in the field, it needs to be applied at a small scale, and most of the time, it is required to be used on a large scale [5,6]. This is inconvenient due to the fact that all these tests and measurements are time-consuming, impractical, and not cost-effective [7,8].

In order to solve the inconveniences mentioned above, a great number of studies about pedotransfer functions (PTFs) were published [7–10]. These mathematical models allow us to estimate the K_S from some soil characteristics, such as texture, field capacity, the permanent wilting point, bulk density, porosity, and organic matter, among others [7–10]. The robustness of the model is linked to the number of physical parameters used to calculate the saturated hydraulic conductivity; the more parameters, the more accurate the prediction. However, as it was mentioned before, depending on the measurements, the PTFs are difficult to get, due to economic resources and the time it takes to measure all the variables, which presents as a limitation for this kind of function and the predictive capacity. Besides, some works have been questioned because the soil in which they want to apply are different from the soil used for their development, such as [11].

In recent years, another alternative has been explored—Artificial Neural Networks (ANNs), which have become a common tool used as a special class of PTFs, for example, [12,13]. ANNs are an artificial intelligence that simulate the behavior of the human brain, and their structures consist of a number of interconnected elements called neurons which are logically arranged in layers, known as input, output, and hidden (see, e.g., [12] and references therein). Each neuron connects to all the neurons in the next layer via weighted connections. In Figure 1, we show a schematic structure for an ANN.

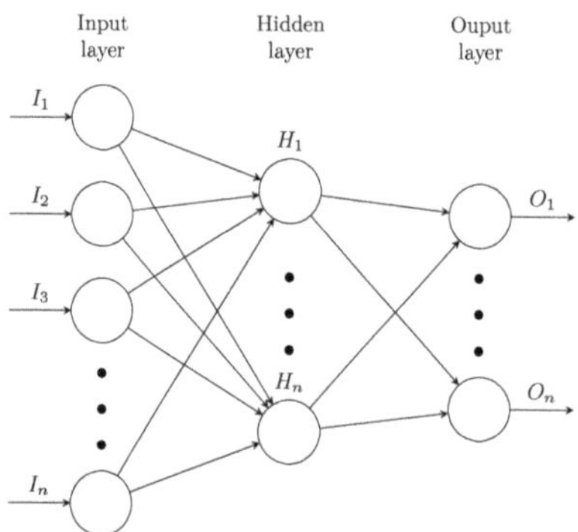

Figure 1. Schematic representation for an ANN structure. Each circle represents a neuron, where the I_j means the neurons in the input layer, and the O_j are the neurons in the output layer. All H_j circles are the neurons in the hidden layers. The arrows represent the weighted connections.

Until now, there has been no analytical way to obtain the ideal network structure (number of hidden layers and neurons inside them) as a function of the complexity of the problem. The structure must be selected by performing a trial-and-error process. ANNs with one or two hidden layers and an adequate number of hidden neurons are found to be sufficient for most problems (e.g., [12,14]). There are also several works studying the ideal number of neurons in the hidden layers [15,16], but these methods present general guidelines for the selection in the number of neurons only.

The ANNs' name comes not only from of their structure, but because they "learn". The most common algorithm used in ANNs for the learning process is back-propagation (e.g., [17,18]). Each neuron belonging to a layer receives weighted inputs from the neurons in the previous layer and processes them to transmit its output to the neurons in the next layer, and this is done through links. Each link is assigned a weight that is no more than a numerical estimate of the strength of the connection. This weighted sum of inputs in a neuron are converted into the numerical estimate that we see, according to the nonlinear transfer function (the most commonly used is the sigmoid function). The ANNs then modify the weights of neurons in response to errors between the actual output values and the target output values, using what is known as gradient descent [19,20]. This is then applied on the sum of the squares of the errors for all the training patterns, until the mean error of the sum squared of all the training patterns is minimal or within the tolerance specified for the problem.

In this work, we use ANNs to obtain the K_S in Irrigation District 023 placed in Queretaro Mexico using a sample of 900 plots. The sample, ANN configurations, and validation tests are described in Section 2. Finally, in Section 3, we show the results, and a comparison between the several configurations obtained in this work and comparisons of our results with PTFs and other ANNs in the literature.

2. Materials and Methods

2.1. Study Area

The Irrigation District 023 is located in Queretaro, Mexico, at 20°18′ to 20°34′ N, 99°56′ to 100°12′ W with an altitude of 1892 M.A.S.L, and it has an area of 11 048 ha. It includes the municipalities of San Juan del Rio and Pedro Escobedo (Figure 2). Its predominant climate is semiarid with summer rains, with an annual precipitation avarage of 599 mm and annual average temperature of 20 °C [21]. The water is conducted through open channels. The main channels are lined with concrete, but all the lateral channels that carry water to the plots are unlined. The separation of the plots in some cases are by trees, unlined channels, drains, or roads [22].

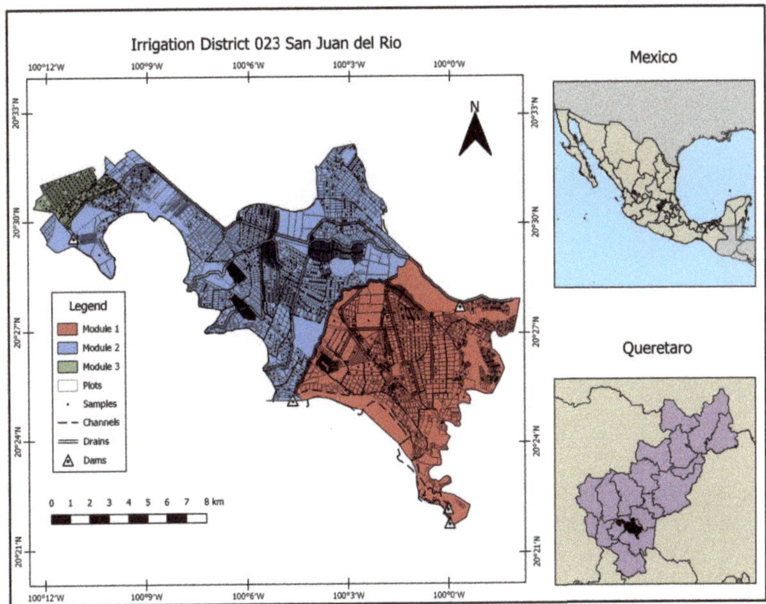

Figure 2. Location map of the sampling points in Irrigation District 023 San Juan del Río, Querétaro.

2.2. Soil Database

An extensive and detailed database description can be found at [23]. In summary, the database used in this study was developed from samplings in 900 plots at Irrigation District 023, San Juan del Rio Queretaro. These samples were sent to the laboratory to obtain the following parameters: soil texture by the Bouyucos hydrometer, bulk density (ρ_a) by the cylinder method of known volume, moisture content at saturation (θ_S), field capacity (FC) and permanent wilting point (PWP) by the method of the pressure membrane pot, and the saturated hydraulic conductivity by the variable head permeameter method. From these measurements, we took seven and considered them through the paper as the input data: percentage of clay, sand and silt, bulk density, permanent wilting point, moisture content at saturation, and field capacity (Table 1).

2.3. The ANNs' Setup

As mentioned in Section 1, there is no analytical way to define the ANN structure, but based on similar several works (e.g., [12,14,16]) we noticed that the ANNs have one or two hidden layers only. With this in mind, we tested several structures for our ANNs with two hidden layers and an additional one with three hidden layers as a test. Details about the ANNs' settings are described below.

Table 1. Statistical properties of data measured in the laboratory.

Variable	Min	Max	Median	Mean	SD	Q1	Q3
Sand (%)	0.07	77.83	28.35	31.14	20.22	13.75	52.00
Clay (%)	2.12	59.46	21.74	21.95	12.06	13.44	30.00
Silt (%)	0.80	92.00	45.27	46.91	23.48	27.30	59.79
ρ_a (g/cm^3)	1.18	1.70	1.40	1.41	0.11	1.32	1.47
PWP (cm^3/cm^3)	0.07	0.35	0.13	0.15	0.05	0.10	0.17
θ_S (cm^3/cm^3)	0.35	0.56	0.47	0.47	0.04	0.45	0.50
FC (cm^3/cm^3)	0.17	0.47	0.29	0.30	0.06	0.25	0.32
K_S (cm/h)	0.05	5.15	0.78	1.42	1.42	0.40	1.80

We began with all the input data (seven) and tested several configurations as follows: we start with two neurons in the first layer and two neurons in the second. Then we continue to vary the number of neurons in the first layer from two to ten, and leave the second layer constant. Next, the number of neurons in the second layer is increased to three and we vary the number of neurons in the first, again, from two to ten, and so on until the number of the second layer varies from two to ten, just like the first one. This input layer has the seven input data mentioned before. Finally, the output layer contains the K_S predicted value.

Then, we changed the number of data in the input layer (decreasing it by one) and repeated the process as explained before. The choice of which parameter has to be removed is based on the importance plot (details in Section 2.4), except for the last configuration, where we remove θ_S instead of the percentage of silt, because θ_S is closely related with the clay percentage. We kept removing input data until we had three measurements only. For each configuration, we varied the number of neurons in the hidden layers from two to ten.

The ANN was programmed using the neuralnet package [24] and the caret package [25], both of them provided by the R software [26].

The neuralnet package trains neural networks using backpropagation, resilient back-propagation (RPROP) with [27] or without weight backtracking [28], or the modified glob-ally convergent version (GRPROP) by [29]. The function allows flexible settings through custom-choice of error and activation function. The caret package (short for Classification And REgression Training) is a set of functions that attempts to streamline the process for creating predictive models. There are many different modeling functions in R. Some have different syntax for model training and/or prediction. The package started off as a way to provide a uniform interface for the functions themselves, as well as a way to standardize common tasks (such as parameter tuning and variable importance). Specifically, we used the train function of this package to perform the cross-validation process.

Furthermore, the calculation of generalized weights [30] is implemented. In this work, we use RPROP as an algorithm type to calculate the neural network, the sum of squared errors as a differentiable function for the calculation of the error and a logistic differentiable function for smoothing the result of the cross-product of the covariate or neurons and the weights.

Using the RMSE measurement, we selected the optimal ANN structure in each case, and finally, we kept this last ANN structure as ideal. The Mean Absolute Error (MAE) measurement was calculated as:

$$\text{MAE} = \frac{1}{n} \sum_{i=1}^{n} |y_i - x_i|,$$

(1)

where y_i are the measurement values, x_i are the predicted values, and n is the total mea-surement. We also calculate the RMSE as:

$$\text{RMSE} = \sqrt{\frac{1}{n} \sum_{i=1}^{n} (y_i - x_i)^2}.$$

(2)

2.4. Cross-Validation

In order to validate our ANN results, we made a cross-validation analysis using the train function from the caret package. This function sets up a grid of tuning parameters for a number of classification and regression routines, fits each model, and calculates a resampling-based performance measure. In this case, we use the parameters optimized to neuralnet, and we generated 25 bootstrap replications for each ANN configuration. Finally, for the prediction of new samples, we used the predict function.

The train function returned several results: the ideal ANN configuration, the best RMSE, MAE and R^2 values for each tested configuration, a RMSE matrix, and an importance plot. This last plot indicates which parameter contributed the most to the K_S final approximation. Based on this last plot, we decided which input data would be removed on the next run, with the exception already mentioned above.

3. Results and Discussion

As mentioned in Section 2, we generated several ANNs which contained all the combinations from two to ten neurons in each hidden layer. In Figure 3 we show the results for these tests. In the x axis we have the number of neurons in the first hidden layer, the y axis represents the RMSE value obtained from the 25 bootstrap replications, and each color represents the number of neurons in the second hidden layer. The election of the best configuration is based on the smallest RMSE value in these plots. Additionally, from the top to the bottom and from the left to the right, we show the variation in the input data. Another result is shown in Figure 4. This plot is a 10×10 matrix where the color represents the RMSE value for each configuration, the rows are the number of neurons in the second hidden layer, and the columns are the number of neurons in first hidden layer. Remember that each plot differs from the other in the number of input data in the same way that Figure 3. In general, we can see that a small number of neurons in the first layer presents higher values of RMSE. Another important result is presented in the Figure 5 where we show a density plot representation (varying the input data from top to bottom) for RMSE, MAE, and R^2 measurements derived from the bootstrap analysis. From these plots we have small variations for each measurement which indicate that the results are not highly dependent on the ANN configurations. Finally, in Figure 6, we show the importance plots, which are described in the cross-validation section presented previously. These plots help us to decide which parameter we must keep or eliminate in each run when we have different numbers of input data.

In order to explore another possibility and to get more confident results, we made a test increasing the number of hidden layers to three. We applied the same process explained previously for this new configuration, and we got an improvement of ~9% in the RMSE values, but with 15 times more computation time. This was the reason for dismissal of these configurations.

In Table 2 we present the three best configurations of each run, where we have the number of input data, the ANN structure, and the RMSE, MAE, and R^2 measurements.

Following the results, we present Figure 7, where we compare the K_S measurements with the K_S predicted by the ANNs for all tested configurations. The dotted line is a 1:1 relation and the R^2 is obtained from the train function. The plotted values were obtained applying the best ANN configuration for each run (Figure 7). The histograms in the top-left corner show the residuals (ΔK_S), which is the difference between the predicted and measured values.

Recall that to obtain the ideal neuron configuration, we apply the train package, which allows us to obtain the RMSE, MAE, and R^2 data for each arrangement of neurons in the hidden layers. The only variation was the input data (ranging from 7 to 3). Therefore, it was possible to obtain five different boxplots for the RMSE, whose only difference would be the aforementioned input data. This is shown in Figure 8, where a trend is observed for the RMSE to fall, while the number of input data decreases.

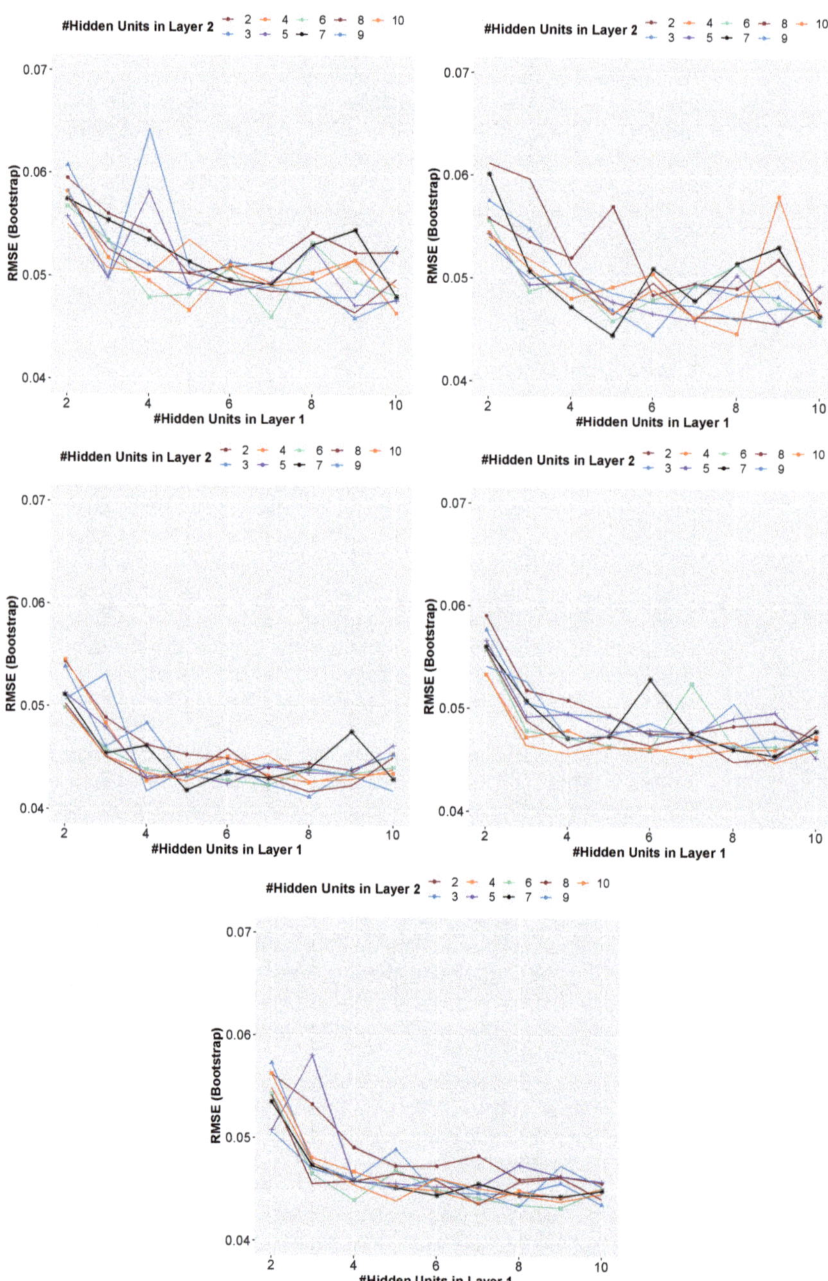

Figure 3. ANN test for choosing the ideal number of hidden neurons varying the input data. From top to bottom and left to right the number of input data goes from 7 to 3. The x axis represents the number of neurons in the first hidden layer and the y axis is the RMSE value. Each color is a different number of neurons in the second hidden layer.

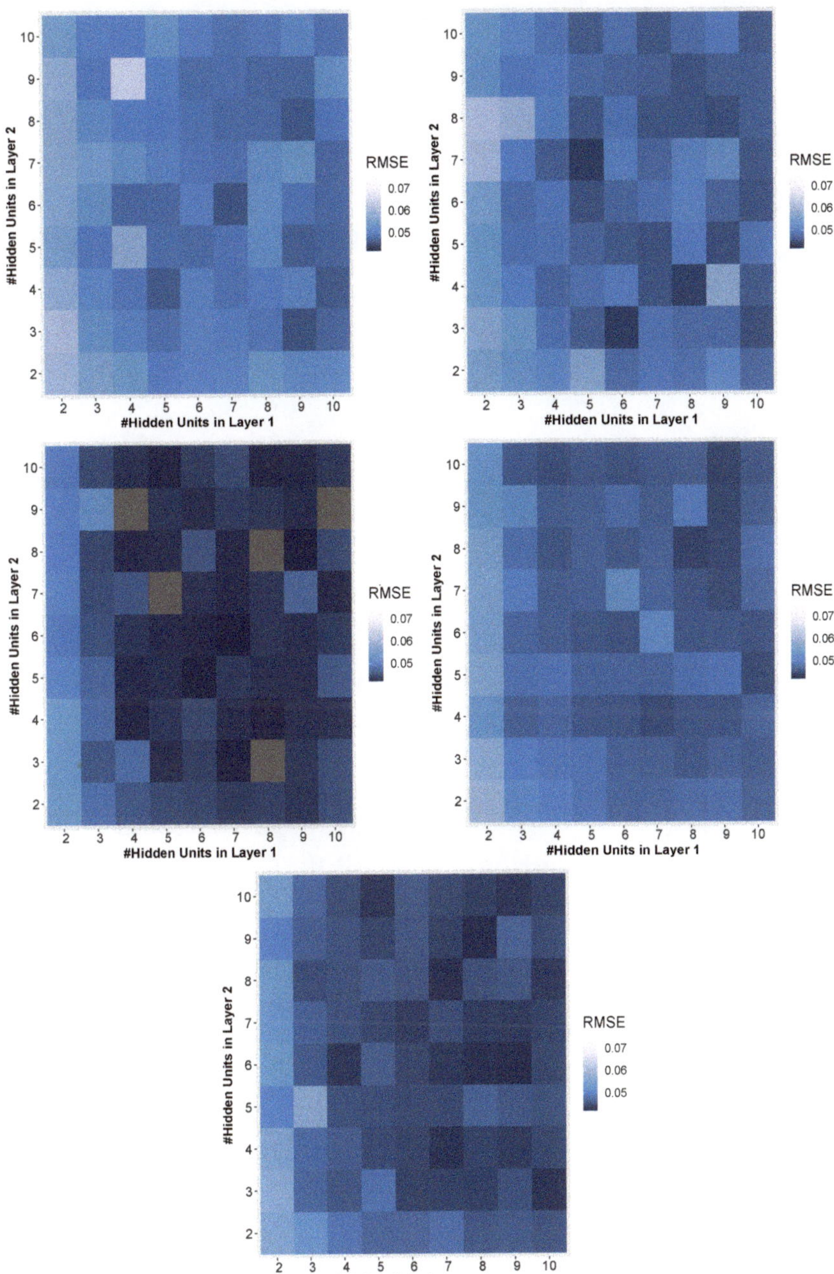

Figure 4. RMSE matrix for different number of neurons in the hidden layers. From top to bottom and left to right, the number of input data goes from 7 to 3. The x axis represents the number of neurons in the first hidden layer and the y axis is the number of neurons in the second hidden layer.

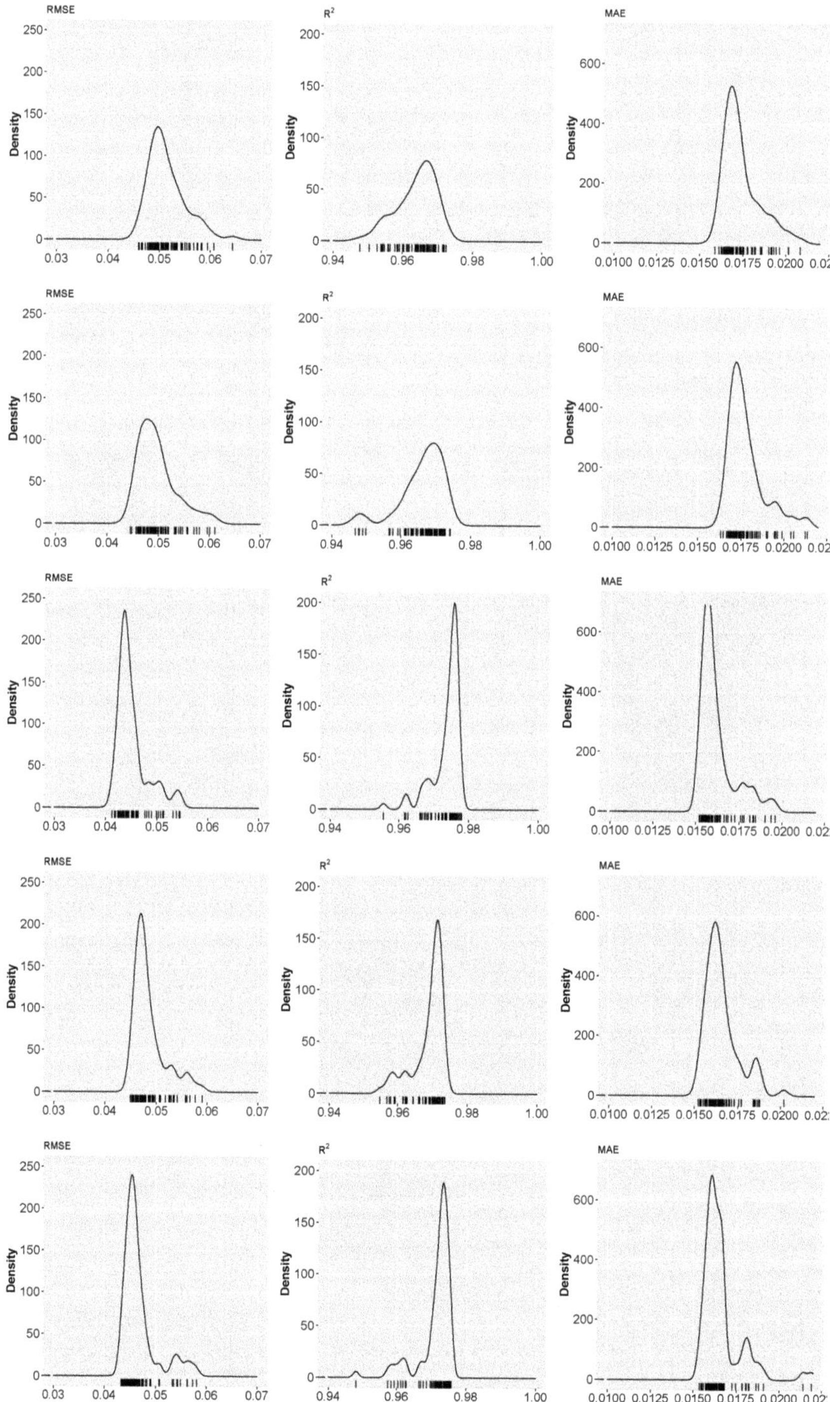

Figure 5. Density plots for (from left to right) RMSE, R^2, and MAE resulting from the 25 bootstrap replications. From top to bottom, the number of input data goes from 7 to 3.

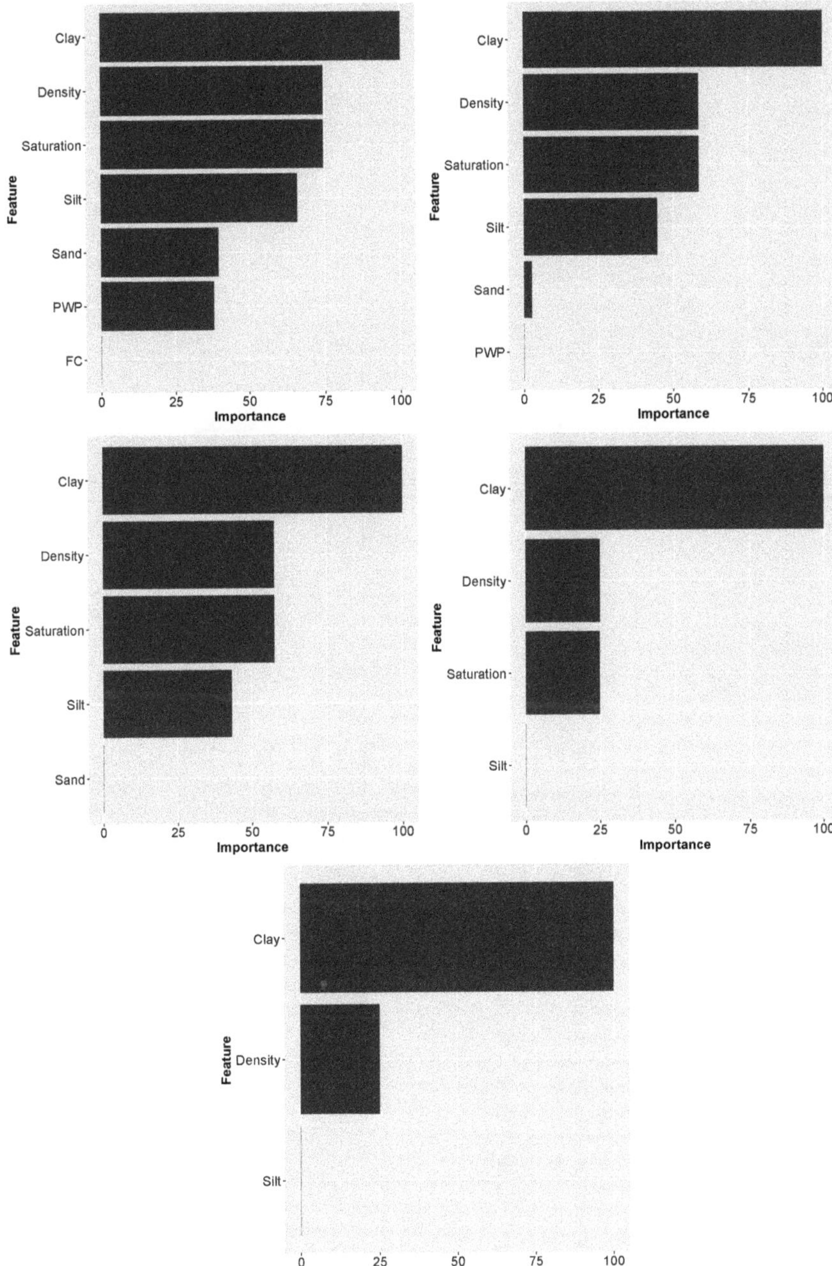

Figure 6. Importance plots referring to the weight of each variable in the calculations. From top to bottom and from right to left, the number of input data goes from 7 to 3.

Table 2. The top three configurations for ANN structure and their statistical measurements. (1) # Input data (2) contains the ANN neurons' structure (Input-Hidden1-Hidden2-Ouput), where each number represents the quantity of neurons used in each layer, (3) the RMSE measurements, (4) the MAE measurements, and (5) the R^2 measurements.

# Input Data	ANN Structure	RMSE (cm/h)	MAE (cm/h)	R^2
7	**7-9-3-1**	**0.0459**	**0.0159**	**0.9725**
	7-7-6-1	0.0460	0.0164	0.9720
	7-10-4-1	0.0465	0.0162	0.9715
6	**6-5-7-1**	**0.0445**	**0.0171**	**0.9740**
	6-6-3-1	0.0455	0.0171	0.9742
	6-8-4-1	0.0447	0.0163	0.9739
5	**5-8-3-1**	**0.0413**	**0.0152**	**0.9780**
	5-4-9-1	0.0417	0.0156	0.9774
	5-8-8-1	0.0418	0.0152	0.9777
4	**4-9-10-1**	**0.0449**	**0.0152**	**0.9736**
	4-8-8-1	0.0450	0.0156	0.9735
	4-9-9-1	0.0452	0.0155	0.9734
3	**3-9-6-1**	**0.0433**	**0.0155**	**0.9757**
	3-8-9-1	0.0434	0.0154	0.9757
	3-8-6-1	0.0436	0.0160	0.9755

In Table 3 we show the results obtained from the literature with PTFs or ANNs and compare them with this work.

Table 3. Comparison between several works for obtaining K_S.

Model	RMSE	R^2	Type
This work	**0.0413**	**0.9780**	**ANN**
Tamari et al. [31]	0.0707	NA	ANN
Brakensiek et al. [9]	0.1370	0.9953	PTF
Erzin et al. [12]	0.1700	0.9970	ANN
Saxton et al. [32]	0.1895	0.9915	PTF
Parasuraman et al. [33]	0.1900	NA	ANN
Trejo-Alonso et al. [23]	0.1983	0.9901	PTF
Cosby et al. [34]	0.4325	0.9546	PTF
Ahuja et al. [35]	0.6498	0.8910	PTF
Schaap & Leij [36]	0.7130	NA	ANN
Vereecken et al. [37]	0.7143	0.9307	PTF
Minasny et al. [38]	0.7330	NA	ANN
Ferrer-Julià et al. [39]	1.3018	0.4083	PTF
Merdun et al. [40]	3.5110	0.5240	ANN

The results for RMSE obtained in this work are better in, at least, 35% compared with the ones presented by [31], and we reported the fifth-best value for R^2.

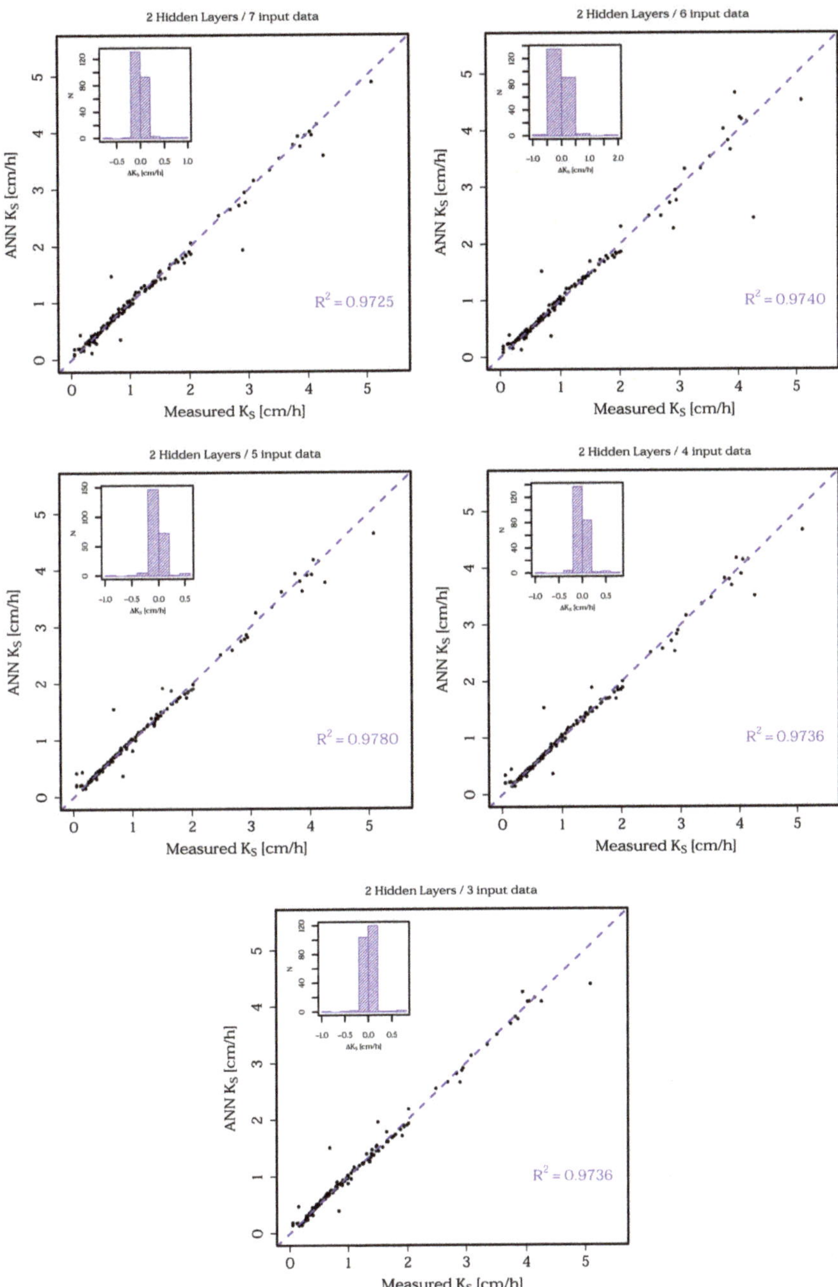

Figure 7. Plots for the comparison between K_S measurement in the field with the K_S value obtained applying the different ANN configurations. The dotted line is a 1:1 desirable relation. The histograms represent the residual distribution (ΔK_S).

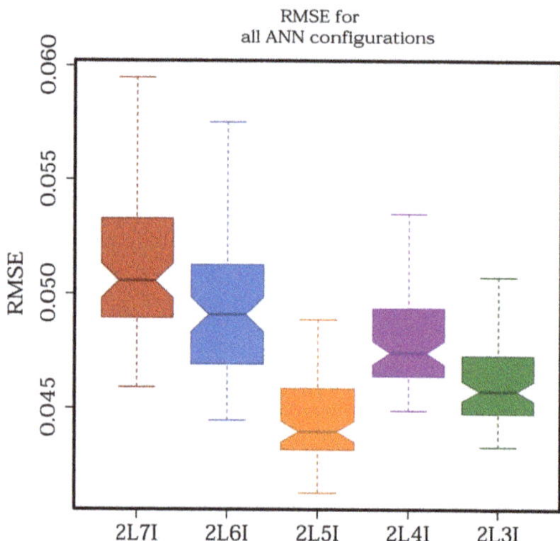

Figure 8. Boxplot for each run of the train package. The names means two hidden layers and the number of input data used in each run.

Another advantage of ANN is that the initial shape for the function to get the relation between the variables or a principal component analysis is not needed. Additionally, as we can see in Figure 3 and Table 2, the results are independent of the ANN structure. This is supported by the fact that the RMSE values for different configurations are very similar (the largest difference is ∼10%), and for R^2 values the difference is ∼5%, and we got ∼11% for MAE. Besides, the Figure 5 presents an almost gaussian distribution for these three statistical measurements, which is in agreement with a non-biased result. In this Figure, we also see that the density distributions are narrower, while the number of input data is smaller. This tells us that in contrast with the PTF models, we found a tendency which indicates that the less input data we have, the more accurate our prediction of K_S is, as well as the Figure 8 shown this tendency too. This result can be explained by the following reasons. Based on the Principal Component Analysis of [23], we noted that the principal variables contributing the most to the sample were K_S, the percentage of clay, θ_S, PWP, and FC, which is supported by the importance plots (Figure 6). Additionally, for the 900 analyzed samples contained in 10 of the 12 existing types of soil (according to the USDA Textural Soil Classification), they showed that the infiltration rate depended directly on the percentage of clay and the ρ_a.

4. Conclusions

In this work, we developed five Artificial Neural Networks in order to calculate the saturated hydraulic conductivity based on the sample used by [23]. All networks consist of one input layer, two hidden layers, and one output layer. We tested a network with three hidden layers, but with little better results. We took 75% of the sample for training and 25% for validation. We also tested all the possible combinations for the number of neurons in each hidden layer, taking into account that the number of neurons for each hidden layer will vary from 2 to 10 neurons. Finally, we selected the best number of neurons in each layer based on RMSE measurements obtained from a cross-validation analysis.

The results show that, compared with other works, we get better or similar results for RMSE and R^2 measurements and similar configurations for our ANN. Finally, we can say that if the necessary resources are available to obtain a large number of data in the field, it is necessary to develop a study of PTF as well as ANN to compare the results of each process and be able to choose the best option between both of them. The latter will not only be based on the RMSE or R^2 measurements, but also on the desired application (a statistical ground property study or prediction for irrigation proposes). A more detailed study to define an exact range of the amount of data needed from a reliable artificial neural network study should be carried out, but the latter is beyond the objective of this work. Besides, we have to be more careful in the characteristics of the sample where the models come out. In our case, we analyzed 10 of the 12 types of soils where the bulk density and the percentage of clay became more important parameters compared to others. This made our models more reliable for almost any type of soil.

The coupling of the Saint Venant and Richards equations is the complete mathematical model for modeling gravity irrigation [41]. However, its use requires detailed information on the physical properties of the soil, as well as a series of field and laboratory experiments that can be expensive [42,43]. In this way, the results used in this article, combined with some rapid field and laboratory tests, can be an excellent alternative to reduce costs and the time used to obtain that information.

Finally, the application of artificial neural networks have been demonstrated to successfully solve classification and prediction problems, and this is probably for the nonlinear relation between the variables. The calculation of the saturated hydraulic conductivity in this work proves that we need only three variables to predict new values, but the soil properties are crucial for the correct application of these models in contrast with the ANN configuration, which has been proved to play a minor role in the final results.

Author Contributions: Conceptualization, C.C. and C.F.; methodology, C.C., J.T.-A. and A.Q.; software, J.T.-A. and B.G.-C.; validation, B.G.-C., J.T.-A. and A.G.-L.; formal analysis, C.C. and J.T.-A.; investigation, A.G.-L. and J.T.-A.; resources, C.C., C.F. and A.Q.; data curation, C.C. and A.Q.; writing—original draft preparation, J.T.-A.; writing—review and editing, C.C., C.F., A.Q. and A.G.-L.; visualization, J.T.-A. and B.G.-C.; supervision, C.C., C.F. and A.Q.; project administration, C.C.; funding acquisition, C.C. All authors have read and agreed to the published version of the manuscript.

Funding: This research was supported as part of a collaboration between the National Water Commission (CONAGUA, according to its Spanish acronym); The Water Users Association of Second Unit of Module Two, Irrigation District No. 023; the Irrigation District 023, San Juan del Río, Querétaro; and the Autonomous University of Querétaro, under the program RIGRAT 2015–2019.

Conflicts of Interest: The authors declare no conflict of interest regarding the publication of this paper.

References

1. Fuentes, S.; Trejo-Alonso, J.; Quevedo, A.; Fuentes, C.; Chávez, C. Modeling Soil Water Redistribution under Gravity Irrigation with the Richards Equation. *Mathematics* **2020**, *8*, 1581. [CrossRef]
2. Chávez, C.; Fuentes, C. Design and evaluation of surface irrigation systems applying an analytical formula in the irrigation district 085, La Begoña, Mexico. *Agric. Water Manag.* **2019**, *221*, 279–285. [CrossRef]
3. Di, W.; Xue, J.; Bo, X.; Meng, W.; Wu, Y.; Du, T. Simulation of irrigation uniformity and optimization of irrigation technical parameters based on the SIRMOD model under alternate furrow irrigation. *Irrig. Drain* **2017**, *66*, 478–491.
4. Gillies, M.H.; Smith, R.J. SISCO: Surface irrigation simulation, calibration and optimization. *Irrig. Sci.* **2015**, *33*, 339–355. [CrossRef]
5. Saucedo, H.; Zavala, M.; Fuentes, C. Complete hydrodynamic model for border irrigation. *Water Technol. Sci.* **2011**, *2*, 23–38.
6. Weibo, N.; Ma, X.; Fei, L. Evaluation of infiltration models and variability of soil infiltration properties at multiple scales. *Irrig. Drain* **2017**, *66*, 589–599.
7. Zhang, Y.; Schaap, M.G. Estimation of saturated hydraulic conductivity with pedotransfer functions: A review. *J. Hydrol.* **2019**, *575*, 1011–1030. [CrossRef]
8. Abdelbaki, A.M. Evaluation of pedotransfer functions for predicting soil bulk density for U.S. soils. *Ain Shams Eng. J.* **2018**, *9*, 1611–1619. [CrossRef]
9. Brakensiek, D.; Rawls, W.J.; Stephenson, G.R. *Modifying SCS Hydrologic Soil Groups and Curve Numbers for Rangeland Soils*; Paper No. PNR-84203; ASAE: St. Joseph, MN, USA, 1984.
10. Rasoulzadeh, A. Estimating Hydraulic Conductivity Using Pedotransfer Functions. In *Hydraulic Conductivity-Issues, Determination and Applications*; Elango, L., Ed.; InTech: Rijeka, Croatia, 2011; pp. 145–164.
11. Moreira, L.; Righetto, A.M.; Medeiros, V.M. Soil hydraulics properties estimation by using pedotransfer functions in a northeastern semiarid zone catchment, Brazil. International Environmental Modelling and Software Society, 2004, Osnabrueck. Complexity and Integrated Resources Management. In *Transactions of the 2nd Biennial Meeting of the International Environmental Modelling and*

Software Society, iEMSs 2004; International Environmental Modelling and Software Society: Manno, Switzerland, 2004; Volume 2, pp. 990–995.

12. Erzin, Y.; Gumaste, S.D.; Gupta, A.K.; Singh, D.N. Artificial neural network (ANN) models for determining hydraulic conductivity of compacted fine-grained soils. *Can. Geotech. J.* **2009**, *46*, 955–968. [CrossRef]
13. Agyare, W.A.; Park, S.J.; Vlek, P.L.G. Artificial Neural Network Estimation of Saturated Hydraulic Conductivity. *VZJAAB* **2007**, *6*, 423–431. [CrossRef]
14. Sonmez, H.; Gokceoglu, C.; Nefeslioglu, H.A.; Kayabasi, A. Estimation of rock modulus: For intact rocks with an artificial neural network and for rock masses with a new empirical equation. *Int. J. Rock Mech. Min. Sci. Geomech. Abstr.* **2006**, *43*, 224–235. [CrossRef]
15. Grima, M.A.; Babuska, R. Fuzzy model for the prediction of unconfined compressive strength of rock samples. *Int. J. Rock Mech. Min. Sci. Geomech. Abstr.* **1999**, *36*, 339–349. [CrossRef]
16. Haque, M.E.; Sudhakar, K.V. ANN back-propagation prediction model for fracture toughness in microalloy steel. *Int. J. Fatigue* **2002**, *24*, 1003–1010. [CrossRef]
17. Singh, T.N.; Gupta, A.R.; Sain, R. A comparative analysis of cognitive systems for the prediction of drillability of rocks and wear factor. *Geotech. Geol. Eng.* **2006**, *24*, 299–312. [CrossRef]
18. Erzin, Y.; Rao, B.H.; Singh, D.N. Artificial neural network models for predicting soil thermal resistivity. *Int. J. Therm. Sci.* **2008**, *47*, 1347–1358. [CrossRef]
19. Rumelhart, D.H.; Hinton, G.E.; Williams, R.J. Learning internal representation by error propagation. In *Parallel Distributed Processing*; Rumelhart, D.E., McClelland, J.L., Eds.; MIT Press: Cambridge, MA, USA, 1986; Volume 1, Chapter 8.
20. Goh, A.T.C. Back-propagation neural networks for modelling complex systems. *Artif. Intel. Eng.* **1995**, *9*, 143–151. [CrossRef]
21. Comisión Nacional del Agua (CONAGUA). *Estadísticas del Agua en México*; CONAGUA: Coyoacán, México, 2018; p. 306.
22. Chávez, C.; Limón-Jiménez, I.; Espinoza-Alcántara, B.; López-Hernández, J.A.; Bárcenas-Ferruzca, E.; Trejo-Alonso, J. Water-Use Efficiency and Productivity Improvements in Surface Irrigation Systems. *Agronomy* **2020**, *10*, 1759. [CrossRef]
23. Trejo-Alonso, J.; Quevedo, A.; Fuentes, C.; Chávez, C. Evaluation and Development of Pedotransfer Functions for Predicting Saturated Hydraulic Conductivity for Mexican Soils. *Agronomy* **2020**, *10*, 1516. [CrossRef]
24. Fritsch, S.; Guenther, F; Wright, M.N. Neuralnet: Training of Neural Networks. R Package Version 1.44.2. Available online: https://CRAN.R-project.org/package=neuralnet (accessed on 19 October 2020)
25. Kuhn, M. Caret: Classification and Regression Training. R Package Version 6.0-86. 2020. Available online: https://CRAN.R-project.org/package=caret (accessed on 22 January 2021).
26. R Core Team. *R: A Language and Environment for Statistical Computing*; R Foundation for Statistical Computing: Vienna, Austria, 2020. Available online: https://www.R-project.org/ (accessed on 12 January 2021).
27. Riedmiller, M. *Rprop—Description and Implementation Details*; Technical Report; University of Karlsruhe: Karlsruhe, Germany, 1994.
28. Riedmiller, M.; Braun, H. A direct adaptive method for faster backpropagation learning: The RPROP algorithm. In Proceedings of the IEEE International Conference on Neural Networks (ICNN), San Francisco, CA, USA, 28 March–1 April 1993; pp. 586–591.
29. Anastasiadis, A.; Magoulas, G.D.; Vrahatis, M.N. New globally convergent training scheme based on the resilient propagation algorithm. *Neurocomputing* **2005**, *64*, 253–270. [CrossRef]
30. Intrator, O.; Intrator, N. Using Neural Nets for Interpretation of Nonlinear Models. In *Proceedings of the Statistical Computing Section*; American Statistical Society: San Francisco, CA, USA, 1993; pp. 244–249.
31. Tamari, S.; Wösten, J.H.M.; Ruiz-Suárez, J.C. Testing an Artificial Neural Network for Predicting Soil Hydraulic Conductivity. *Soil Sci. Soc. Am. J.* **1996**, *60*, 1732–1741. [CrossRef]
32. Saxton, K.; Rawls, W.J.; Romberger, J.S.; Papendick, R.I. Estimating generalized soil water characteristics from texture. *Soil Sci. Soc. Am. J.* **1986**, *5*, 1031–1036. [CrossRef]
33. Parasuraman, K.; Elshorbagy, A.; Si, B. Estimating Saturated Hydraulic Conductivity In Spatially Variable Fields Using Neural Network Ensembles. *Soil Sci. Soc. Am. J.* **2006**, *70*, 1851–1859. [CrossRef]
34. Cosby, B.; Hornberger, G.; Clapp, R.; Ginn, T. A statistical exploration of the relationship of soil moisture characteristics to the physical properties of soils. *Water Resour. Res.* **1984**, *20*, 682–690 [CrossRef]
35. Ahuja, L.R.; Naney, J.W.; Green, R.E.; Nielsen, D.R. Macroporosity to characterize spatial variability of hydraulic conductivity and effects of land management. *Soil Sci. Soc. Am. J.* **1984**, *48*, 699–702. [CrossRef]
36. Schaap, M.G.; Leij, F.J. Using neural networks to predict soil water retention and soil hydraulic conductivity. *Soil Tillage Res.* **1998**, *47*, 37–42. [CrossRef]
37. Vereecken, H.; Schnepf, A.; Hopmans, J.W.; Javaux, M.; Or, D.; Roose, T.; Vanderborght, J.; Young, M.H.; Amelung, W.; Aitkenhead, M.; et al. Modeling Soil Processes: Review, Key Challenges, and New Perspectives. *Vadose Zone J.* **2016**, *15*. [CrossRef]
38. Minasny, B.; Hopmans, J.; Harter, T.; Eching, S.; Tuli, A.; Denton, M. Neural Networks Prediction of Soil Hydraulic Functions for Alluvial Soils Using Multistep Outflow Data. *Soil Sci. Soc. Am. J.* **2004**, *68*, 417–429. [CrossRef]
39. Ferrer-Julià, M.; Estrela-Monreal, T.; Sánchez-del Corral-Jiménez, A.; García-Meléndez, E. Constructing a saturated hydraulic conductivity map of spain using pedotransfer functions and spatial prediction. *Geoderma* **2004**, *123*, 275–277. [CrossRef]
40. Merdun, H.; Cinar, O.; Meral, R.; Apan, M. Comparison of artificial neural network and regression pedotransfer functions for prediction of soil water retention and saturated hydraulic conductivity. *Soil Tillage Res.* **2006**, *90*, 108–116. [CrossRef]

41. Richards, L.A. Capillary conduction of liquids through porous mediums. *Physics* **1931**, *1*, 318–333. [CrossRef]
42. Saucedo, H.; Fuentes, C.; Zavala, M. The Saint-Venant and Richards equation system in surface irrigation: (2) Numerical coupling for the advance phase in border irrigation. *Ing. Hidraul. Mexico* **2005**, *20*, 109–119.
43. Fuentes, C.; Chávez, C. Analytic Representation of the Optimal Flow for Gravity Irrigation. *Water* **2020**, *12*, 2710. [CrossRef]

MDPI
St. Alban-Anlage 66
4052 Basel
Switzerland
Tel. +41 61 683 77 34
Fax +41 61 302 89 18
www.mdpi.com

Water Editorial Office
E-mail: water@mdpi.com
www.mdpi.com/journal/water

www.ingramcontent.com/pod-product-compliance
Lightning Source LLC
LaVergne TN
LVHW070912170726
843515LV00005B/762